CRYSTALLINE AND NON-CRYSTALLINE SOLIDS

Edited by **Pietro Mandracci**

Crystalline and Non-crystalline Solids
http://dx.doi.org/10.5772/61501
Edited by Pietro Mandracci

Contributors

Mario Moreno, Chellaiah Arunkumar, Subramoniam Sujatha, Stefano Carossa, Federico Mussano, Tullio Genova, Salvatore Guastella, Maria Giulia Faga, Ryohei Takei, Giovanni Fanchini, Paola Rivolo, Francesca Frascella, Serena Ricciardi, Micaela Castellino, Mihaela Filipescu, Alexandra Palla Papavlu, Maria Dinescu, Elena Konshina

Published by InTech

Janeza Trdine 9, 51000 Rijeka, Croatia

Publishing Process Manager Romina Rovan

Technical Editor SPi Global

Cover InTech Design team

First published July, 2016

Printed in Croatia

Additional hard copies can be obtained from orders@intechopen.com

Crystalline and Non-crystalline Solids, Edited by Pietro Mandracci

p. cm.

Print ISBN 978-953-51-2445-0
Online ISBN 978-953-51-2446-7

For EU product safety concerns:
IN TECH d.o.o., Prolaz Marije Krucifikse Kozulić 3, 51000 Rijeka, Croatia,
info@intechopen.com or visit our website at intechopen.com.

Meet the editor

Pietro Mandracci was born in Torino in 1970. He received his MD in physics from the Torino University in 1996 and his PhD in electron devices from the Trento University in 2001. Since 2004 he has been Assistant Professor at Politecnico di Torino. During his career, he has worked on the simulation of solar cells, the fabrication of UV detectors based on amorphous semiconductors, the design of multilayer structures for the fabrication of optical filters and the synthesis of advanced thin-film materials and nanostructures. His present research interests deal with thin-film technology and nanotechnology, structural, optical and electrical characterization of materials and nanostructures, as well as their application to electronic and photonic devices, micromechanical systems, and biomedicine.

Contents

Preface

This book is focused on the most recent research regarding some of the most important crystalline and non-crystalline materials, which are presently exploited or could be exploited in the near future, for the fabrication of advanced technological devices in several application fields, including microelectronics, photonics, nanotechnology, as well as biotechnology. The structural properties of materials, whether they show a crystalline structure or not, play a fundamental role in determining whether they are suitable for a specific application. Nowadays, a huge amount of different materials are used in the diverse technological fields, and many of them can show several structures depending on the conditions at which they are synthesized. Understanding the relation between the structure of these materials and their properties is a task of paramount importance and is becoming even more important as the technology advances, requiring more demanding performances from materials.

This book is intended as a contribution to the effort to understand the influence that the crystalline or amorphous structure exerts on the properties of materials. To this aim, some of the materials that are most promising for their use in different technological fields have been studied, namely graphene, titanium oxide, several types of functional metal oxides, porphyrinic crystalline solids, plasma deposited polymers, amorphous silicon, as well as hydrogenated amorphous carbon. These materials have been presented by the authors for their use in different applications, including microelectronics, photonics, and biomedicine.

The book is divided in two sections, each one including four chapters: while the first section is devoted to materials that show different types of crystalline structures, the second section is devoted to amorphous materials. The first chapter of the book deals with graphene, which nowadays is probably one of the most studied materials, discussing how electronic, thermal, and optical properties of graphene-based materials depend strongly on the fabrication method used and can be further manipulated through the use of metal nanoparticles deposited on the graphene surface. Under ideal conditions, self-assembled arrays of nanoparticles can be obtained on graphene-based films for use in new types of nanodevices such as evanescent waveguides.

The second chapter reviews the scientific literature with the scope of assessing what is known about the surface micro/nanotopography and the crystallographic microstructure of titanium dental implants. Also, the correlation between these surface features and the biological outcomes in vitro and in vivo is a primary object of the chapter. Based on the results of the most recent studies, the surface of titanium dental implants may be constituted of anatase, rutile, and amorphous phases. Anatase seems to be more present in arc-oxidized implants, alone or with rutile, according to the oxidation conditions. Rutile and amorphous phases are more frequently found in machined, double-etched, sandblasted, and sandblasted-acid etched implants. Particular interest is raised by the possible presence of brookite, which was found on a commercially available sandblasted-acid etched implant. Taking into consideration the variations in the biological activity of these polymorphs, identification of the TiO_2 phases found in the surface layers of implants should be regarded as fundamental not only by researchers but also by manufacturers.

The aim of the third chapter is to show that material processing by laser-based technologies can lead to the growth of multifunctional thin films with potential in a large area of applications. The synthesis of Hf, Ta, Si, and Al metal oxides relies on the use of pulsed laser deposition (PLD) or radio-frequency (RF)-assisted PLD. It is shown that by tuning the deposition parameters, the materials of interest can be synthesized as compact and dense oxide layers. Parameters, such as substrate temperature, oxygen pressure, or laser wavelength, have a critical impact on the crystallinity of the films, as well as on the characteristic functional properties. When high substrate temperatures are involved in the PLD process, these oxide layers have a crystalline structure and smooth surfaces, with potential in antireflective coatings.

The fourth chapter deals with crystal engineering, which is an emerging area of research in material, biological, and pharmaceutical chemistry, that involves synthesis of new materials, analysis of its structure including intermolecular interactions using x-ray crystallography as well as computational methods. The synthetic porphyrins are of widespread attention because of their close resemblance to naturally occurring tetrapyrrolic pigments and they find a variety of material and biological applications. In this book chapter, some recent findings on detailed crystal structure analysis of a few series of fluorinated porphyrins are disclosed, using single crystal XRD as well as computational Hirshfeld surface analysis in order to understand the role of close contacts involving fluorine in the molecular crystal packing.

The fifth chapter deals with many efforts which have been done to chemically functionalize sensors surface to achieve selectivity towards diagnostics targets, such as DNA, RNA fragments, and protein tumoral biomarkers, through the surface immobilization of the related specific receptor. The aim of the chapter is to report on the study and optimization of ultra-thin plasma polymers and co-polymers, obtained from the vapors of acrylic acid containing a carboxylic group and styrene. The obtained plasma polyacrylic acid, polystyrene, and their copolymer are shown to match specific and critical requirements, such as low thickness (~40 nm) and refractive index (~1.5), high surface density of reactive groups (10^{15}–10^{16} COOH/cm^2), bio-antifouling properties where required, reproducibility, and chemical resistance and stability.

The sixth chapter introduces recent research on amorphous silicon photonics. By exploring high-quality silicon thin-film technology, the authors have demonstrated hydrogenated amorphous silicon waveguides with ultra-low loss, vertical interlayer transition devices for cross coupling between vertically stacked optical circuits. These device technologies are promising for three-dimensional photonic integrated circuits integrated in microelectronics chips. A record low loss of 0.6 dB/cm was achieved for a submicron scale single-mode waveguide, and the VIT devices allow low-loss, broadband, and polarization-insensitive operation.

In the seventh chapter, the results of the study of structural features and optical properties of thin films of amorphous hydrogenated carbon films prepared by plasma-activated chemical vapor deposition of various hydrocarbon precursors are reviewed. It is shown that the refractive index of a-C:H films can be changed in the interval 2.35–1.55 by increasing the deposition rate and the choice of the appropriate hydrocarbon precursor. The features of the vibration spectra of the diamond-like and polymer-like films are discussed. The correlations of the structural peculiarities and of the optical absorption edge, gap width, and conductivity as well as the absorption spectra in visible region and the ratio of the fundamental bands in Raman scattering spectra are estimated. Examples of using the optical properties of the a-C:H films are given.

The last chapter is devoted to the study of amorphous, polymorphous, and microcrystalline silicon, deposited by the plasma-enhanced chemical vapor deposition (PECVD) technique at low temperatures. The main deposition parameters that have strong influence on the optical, electrical, and structural properties of the polymorphous and microcrystalline materials have been studied. Results reveal the key deposition conditions to obtain films with optical and electrical characteristics, which are suitable for applications on thin-film solar cells and semiconductor devices.

Dr. Pietro Mandracci
Department of Applied Science and Technology,
Politecnico di Torino,
Torino, Italy

Section 1

Crystalline Solids

Chapter 1

Graphene Thin Films and Graphene Decorated with Metal Nanoparticles

Paul Bazylewski, Arash Akbari-Sharbaf,
Sabastine Ezugwu, Tianhao Ouyang,
Jaewoo Park and Giovanni Fanchini

Additional information is available at the end of the chapter

http://dx.doi.org/10.5772/63279

Abstract

The electronic, thermal, and optical properties of graphene-based materials depend strongly on the fabrication method used and can be further manipulated through the use of metal nanoparticles deposited on the graphene surface. Metals that strongly interact with graphene such as Co and Ni can form strong chemical bonds which may significantly alter the band structure of graphene near the Dirac point. Weakly interacting metals such as Au and Cu can be used to induce shifts in the graphene Fermi energy, resulting in doping without significant alteration to the graphene band structure. The deposition and nucleation conditions such as deposition rate, annealing temperature and time, and annealing atmosphere can be used to control the size and distribution of metal nanoparticles. Under ideal conditions, self-assembled arrays of nanoparticles can be obtained on graphene-based films for use in new types of nanodevices such as evanescent waveguides.

Keywords: graphene, thin films, metal nanoparticles, optical properties, electronic band structure

1. Introduction

Graphene is a single layer of carbon atoms bonded in a hexagonal honeycomb lattice, resulting in a structure with many desirable characteristics that are attractive for several applications. These unique properties include high charge carrier mobility and thermal conductivity, coupled with a large surface area that is ideal for catalysis or sensing applications. However, utilizing graphene effectively in many technological applications depends on fabrication of the appropriate type of graphene-based material ranging from large area single and few-layer graphene sheets to laminate films sintered from many nano- or micro-meter-sized graphene platelets.

High-quality single- and few-layer graphene up to a thickness of ~10 layers can be successfully fabricated using vacuum-based deposition technologies such as chemical vapor deposition (CVD) [1]. CVD graphene can be grown on metal substrates that are lattice matched to the graphene lattice such as Ni(111) and Cu, using a carrier gas usually composed of CH_4 that reacts with the surface under high-temperature conditions to promote graphene growth. Both substrate types can be used to produce high-quality single- and few-layer graphene using CVD, although the specific growth method varies depending on the substrate material. Graphene grown by CVD methods has been shown to have excellent electronic properties and can closely approximate the predicted theoretical performance of graphene sheets. However, high-quality graphene grown by CVD can be costly to produce and is sensitive to contamination and defects, making it difficult to utilize effectively in many device applications. Since the substrates used in graphene growth is not desirable for all device architectures, the graphene must be transferred to another substrate such as Si/SiO_2. This is commonly accomplished using a solution-based method utilizing a sacrificial polymer such as poly-methyl-methacrylate (PMMA) or another cross-linked polymer, but can introduce defects and contamination [1, 2].

To overcome these difficulties, solution-based methods that do not utilize vacuum systems have been developed. These methods concentrate on chemical exfoliation of graphite to produce colloidal dispersions of few-layer graphene platelets. This approach produces an oxidized form of graphene known as graphene oxide (GO) that is dispersible in aqueous solutions due to intercalation of oxygen functional groups within the graphite lattice. This approach is scalable and versatile in terms of chemical functionalization and use with a variety of substrates. However, one significant disadvantage of colloidal dispersions is films of GO are highly insulating due to a high density of oxygen functional groups, and must be reduced in order to recover the desirable properties of graphene [3]. This can be accomplished using chemical reductants or thermal treatment to remove oxygen and restore conductivity. However, the reduction process is intrinsically energetic and unavoidably results in defects in the graphene lattice that negatively impact the electronic properties of the final graphene film.

Recent advances in solution processing of graphite have been focussed on limiting the use of strong acid treatments to control the oxidation, and instead make use of surfactants to aid the exfoliation process. Ribonucleic acid (RNA) has been used by Sharifi *et al.* [4] as an effective aqueous surfactant to exfoliate graphite that is weakly oxidized in comparison to GO before reduction. Weakly oxidized graphene of this type has been shown to be dispersible in water due to RNA absorption (**Figure 1**) and is conductive as-deposited from solution without further treatment to remove oxygen functional groups.

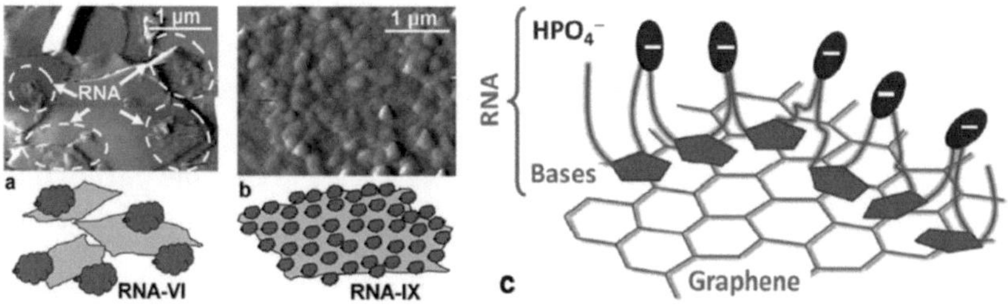

Figure 1. (a, b) Atomic force microscopy phase images demonstrating the arrangement of two types of RNA, RNA VI, and IX aggregates on specific surface regions of small exfoliated graphene flakes. (c) The adhesion mechanism of RNA where a combination of hydrophobic bases and hydrophilic phosphate groups keep the graphene-RNA suspended in water. Reproduced with permission from Ref. [4].

Both types of graphene from either vacuum-based or solution-based fabrication can be utilized as active layers in thin-film electronic devices, or further modified using metal nanoparticles. Doping effects from metal nanoparticles may be used to shift the graphene work function for solar cell applications or light-emitting diode (LED) applications. Organized nanostructures such as self-assembled ordered superlattices of metal nanoparticles can be used as plasmonic waveguides [5, 6]. The electronic band structure of graphene can be altered by applying metallic layers on its surface, where this effect is strongly dependent on the specific type of metal being used. A strong interaction from the formation of strong chemical bonds with metals, such as Co, Ni, and Pd, may significantly alter the band structure of graphene near the Dirac point. For weakly bonding metals (Cu, Al, Ag, Au, and Pt), shifts in the graphene Fermi energy can be induced due to electron transfer, resulting in doping without significant alteration to the graphene band structure.

2. Research and methods

2.1. Graphene thin film fabrication

The use of vacuum techniques to fabricate high-quality large area sheets of single-layer graphene is of great interest for device applications and the study of the fundamental physics of graphene. Metal substrates most commonly used are Ni and Cu because they possess a crystal structure with a lattice spacing well matched to that of graphene. Ni(111) is ideal for graphene growth as it possesses a lattice structure reminiscent of the hexagonal lattice of graphene with similar lattice constants. Graphene is fabricated starting on a polycrystalline Ni substrate that is first annealed in an Ar/H_2 atmosphere at high temperature (800–1000°C) to increase the grain size. A polycrystalline substrate is lower in cost than a single crystal, but intrinsically contains grain boundaries that limit the maximum size of graphene grains. The heated Ni substrate is then exposed to a H_2/CH_4 gas mixture. Upon contact with the Ni, the hydrocarbons decompose and carbon atoms dissolve into the Ni film, forming a solid solution. Cooling of the sample with argon gas causes carbon atoms to diffuse out from the Ni-C solid solution and precipitate on the Ni surface in the form of graphene films. The use of Cu as a substrate is a similar process involving the same carrier gases; however, carbon has a much lower solubility in Cu at elevated temperature than Ni. Since Cu is also well lattice matched to graphene, rather than dissolving, the hydrocarbons decompose on the surface of Cu into a graphene layer [1]. This technique can produce multilayer graphene easily by simply allowing the reaction to proceed for a longer length of time to build up a graphene multilayer. CVD graphene on Cu or Ni can be transferred to other substrates to become part of device architecture or used for further processing with metal nanoparticles or chemical functionalization.

This procedure is advantageous for applications that require single- or few-layer graphene. CVD films are highly transparent (98% transparency to visible light) and conductive (100–1000 Ω/square), making them ideal as a potential replacement for more expensive transparent conducting materials such as indium tin oxide (ITO) in solar cell devices. However, the need for a large, single-crystal substrate limits the ultimate size of a single-grain graphene sheet. The solution transfer process used to integrate CVD graphene into devices further limits the use of CVD graphene by introducing defects and a PMMA residue that is difficult to sufficiently remove. These factors limit the ultimate size of defect-free graphene grains that can be obtained using a CVD graphene fabrication process.

Alternative to vacuum-based techniques, the methods based on chemical exfoliation of graphite can be used to obtain colloidal suspensions of graphene in aqueous or solvent solution. This approach is scalable, has the potential for high-volume production, and is versatile in terms of chemical functionalization. Graphite oxide has been mainly produced by one of three common methods, the Brodie, Staudenmaier, or Hummers methods, which all utilize the oxidation of graphite in the presence of strong acids and oxidants [2]. Brodie and Staudenmaier use a combination of potassium chlorate ($KClO_3$) with nitric acid (HNO_3) to oxidize graphite, and Hummers treats the graphite with potassium permanganate ($KMnO_4$) and sulfuric acid (H_2SO_4). The level of oxidation can be varied on the basis of the specific method and reaction conditions, and the precursor graphite material used. Graphite oxide consists of a layered structure of graphene oxide sheets that are strongly hydrophilic due to an excess of oxygen functional groups such that intercalation of water molecules between the layers readily occurs.

Colloidal solutions of GO can be used to produce laminate films formed of graphene platelets using a variety of methods to separate GO platelets from solution such as vacuum filtration, dip coating, spin coating, or Langmuir-Blodgett film assembly [3]. However, the use of GO directly produces insulating films that must be reduced to remove a fraction of oxygen functional groups and restore conductivity. Several methods exist to reduce GO including chemical reduction using reducing agents such as hydrazine or hydroquinone, thermal annealing, or ultraviolet-assisted reduction. Although successful to reduce GO, defects are unavoidably introduced in the graphene lattice after reduction. Defects act to reduce the electrical and thermal conductivity of reduced GO and can also spread to unzip large GO sheets into smaller domains. Alternative methods exist to produce graphene suspensions using only weak oxidation of graphite combined with another material to behave as a surfactant and promote dispersion in water. Surfactants such as sodium dodecylbenzene sulfonate (SDBS) can be used to enhance graphite exfoliation without the use of strong oxidation treatments to saturate the graphite layers with oxygen functional groups and promote water solubility. RNA has been used as a low cost and biocompatible surfactant to disperse weakly oxidized graphite in water [4]. The use of an additional material to aid in exfoliation allows the graphite to be dispersed in water without being strongly oxidized, and therefore does not require reduction to become conductive.

2.2. Electrical and thermal properties of graphene thin films

Graphene thin films formed from interlocking and overlapping platelets of graphene or GO have electrical, optical, and thermal properties that diverge from those of single-layer graphene. A single layer of graphene has optical transparency of approximately 98%, which decreases as the number of layers increases. The electrical conductivity of few-layer graphene is typically lower than a multilayer, providing a trade-off of increased conductivity for lower transparency. Multilayer graphene laminates, therefore, have lower optical transparency compared to single-layer graphene, but can have much lower resistivity approaching 100–1000 Ω/square.

Graphene and graphene-based thin films, however, are superior to ITO for thermal management applications since the thermal conductivity of ITO is relatively low (5.9 W m^{-1} K^{-1}) compared to graphene laminate films (~10^2 W m^{-1} K^{-1}) [7]. Interestingly, the excellent thermal properties of single-layer graphene are also retained to a large extent in thin graphite multilayers and graphene laminate films. In ITO, the thermal properties at room temperature conditions are determined primarily by the electronic band structure. By contrast, thermal characteristics in graphene-based materials require a more complex interpretation due to important effects from lattice vibrations [7]. The thermal and electrical conductivity of graphene-based thin films strongly depend on the average number of graphene layers forming the platelets, their lateral size, and density of the flakes including voids between neighboring domains [8, 9].

Electrical conductivity of graphene-based films is greatly affected by both the preparation method and the post-treatments. Graphene-based films produced using RNA as a surfactant reveal a trend of decreasing sheet resistance with annealing temperature from Sharifi *et al.* shown in **Figure 2** [4]. The decrease in sheet resistance as a function of the annealing temperature demonstrates that annealing is effective in improving the conductivity of RNA-surfactant-based films without altering transparency. RNA is insulating and tends to decompose above room temperature; the amount of RNA aggregates intercalating the graphene flakes decreases dramatically upon annealing at high temperature (510°C), leading to better conduction between flakes and improved overall conductivity. The effect of thermal annealing is compared to functionalization where the sheet resistance decreased after initial exposure to a nitric acid (HNO_3) bath, followed by a further decrease after soaking in thionyl chloride ($SOCl_2$) without a decrease in transparency. This combination of treatments leads to thin films with sheet resistance of 200 Ω/square at 50% transmittance, or 2.3 MΩ/square at 85% transmittance. Although this performance is inferior to graphene materials epitaxially grown on copper, it is comparable to other classes of graphene-based thin films such as SDBS-based graphene dispersions.

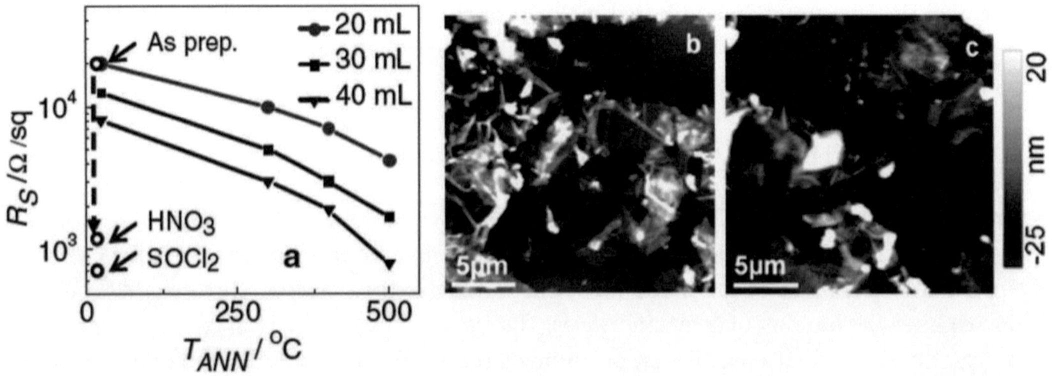

Figure 2. (a) Sheet resistance as a function of annealing temperature for graphene-RNA VI films at different filtration volumes. (b) AFM image before annealing. (c) AFM image of the same sample after annealing. Reproduced with permission from Ref. [4].

2.3. Graphene thin films decorated with metal nanoparticles

Graphene-based film properties can be altered through the use of metal nanoparticles (NPs) applied to the surface. Introducing metal nanoparticles to graphene has been used to shift the work function of graphene, open a band gap, and produce ordered nanostructure super lattices depending on the strength of metal-graphene interaction. For strongly interacting metals, the graphene-metal separation depends on the available bonding locations where the graphene has the appropriate position with respect to the substrate such that bonding can be optimized and the separation is minimized.

Gold nanoparticles are an example of weakly interacting metal particles that can be used to induce plasmonic effects that are useful in applications such as plasmonic solar cells and optical memory devices. Venter *et al.* [10] used graphene thin films obtained from RNA-surfactant-assisted exfoliation of graphite as a substrate that was coated with Au_{144} NP clusters using spin coating from a toluene solution containing $Au_{144}(SCH_2CH_2Ph)_{60}$ molecules synthesized using a modified Brust-Schiffrin method. SEM images of nucleated Au NPs are shown in **Figure 3**. A post-annealing step is used to promote nucleation of Au_{144} into NPs and also releases the capping ligands of Au_{144} molecules, similar to what has been observed for the nucleation of Au_{25} molecular nanoclusters and the expulsion of the corresponding ligands with the formation of Au NPs in polymers [11]. Pre-annealing was also utilized to decompose the RNA and remove it from the film, leaving defects and impurity on the graphene surface that may act as nucleation sites for Au_{144} to aggregate and form NPs when annealed. Similar deposition on graphene-based films using SDBS did not show nucleation of nanoparticles, indicating RNA plays a crucial role in Au NP nucleation. Metal nanoparticles nucleated in this way are advantageous because a uniform layer of Au NPs is formed with a size independent of the number of graphene layers, in contrast to traditional methods of annealing colloidal or thermally evaporated gold thin films which have a size and morphology dependent on the number of graphene layers beneath.

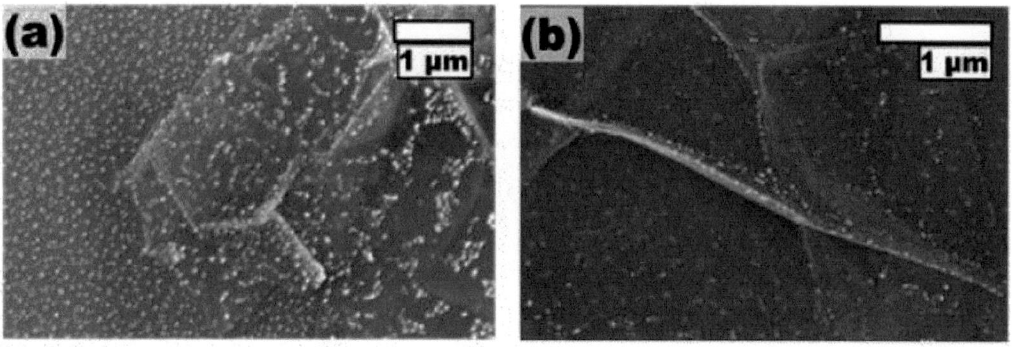

Figure 3. SEM images of graphene thin films deposited on glass decorated with Au NPs. Pre-annealing of the substrate at 300°C for 20 min was combined with 200°C heating for 30 min after Au deposition to nucleate Au-NPs. (a) Au-NPs spin coated at 3000 rpm for 60 s. (b) Au-NPs spin coated at 4000 rpm for 60 s. Reproduced with permission from Ref. [10].

An attractive avenue for producing doped large-area graphene thin films while altering their electronic band structure near the Dirac point in a limited way is through the use of metal NPs. This is particularly true in many cases where assembling metallic layers on top of graphene sheets is not viable for practical applications for which access to the graphene surface is required. **Figures 4a, b** and **c** from Akbari-Sharbaf *et al.* [5] show scanning electron micrographs (SEM) of Cu-NPs assembled on graphene-RNA-based films deposited by vacuum filtration followed by drying in a vacuum desiccator and annealing at 550°C for 5 h to remove RNA. Cu-NPs were deposited using RF sputtering of Cu metal target under high-vacuum conditions using very short sputter times to produce thin metal films, followed by annealing at temperatures ranging from 300–550°C for 1–4 h in a glove box under a nitrogen atmosphere. Annealing under an inert atmosphere was found to be an essential step in obtaining well-isolated Cu-NPs, while non-annealed Cu films formed a semi-continuous system of interconnected Cu particles. **Figures 4d** and **e** show how the average diameter of Cu-NPs and the fraction of graphene area covered can be controlled using the annealing temperature, while a longer annealing time promotes the formation of larger, more isolated NPs.

The work function of graphene-based films decorated with Cu-NPs was further investigated using scanning Kelvin probe force microscopy (SKPFM) shown in **Figure 5b** for a graphene substrate and **Figure 5e** for an ITO substrate. SKPFM is based on small electrostatic forces that are created between a conducting AFM tip and the sample when the two systems are in close proximity to determine the work function of a film. **Figures 5c** and **f** show the work function relative to Cu-NP size, indicating that the graphene-Cu work function decreases with increasing particle size with respect to bare graphene. Graphene-based films contain fewer free electrons compared to ITO, resulting in a larger work-function shift due to electron transfer from Cu-NPs.

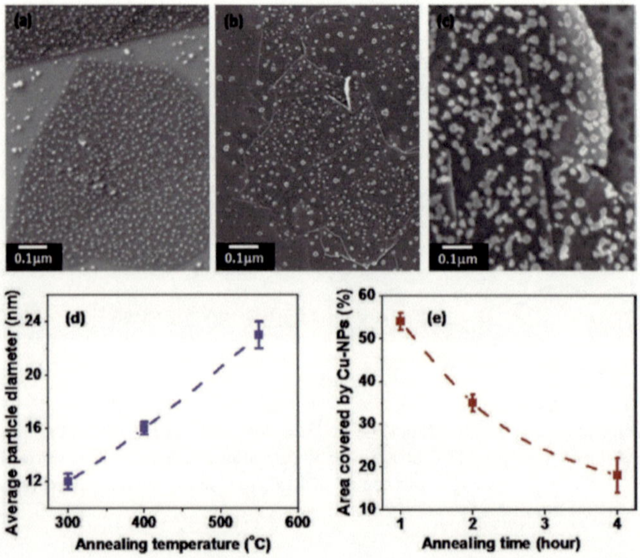

Figure 4. SEM images of graphene-based thin films decorated with Cu-NPs. Cu-NPs were deposited at different sputtering times of 2, 3, and 5 min, respectively, and annealed at (a) 300, (b) 400, and (c) 550°C. (d) Variation of the average diameter of Cu-NPs decorated on the graphene films shown in the SEM images. (e) Fraction of graphene surface covered by Cu-NPs. Reproduced with permission from Ref. [5].

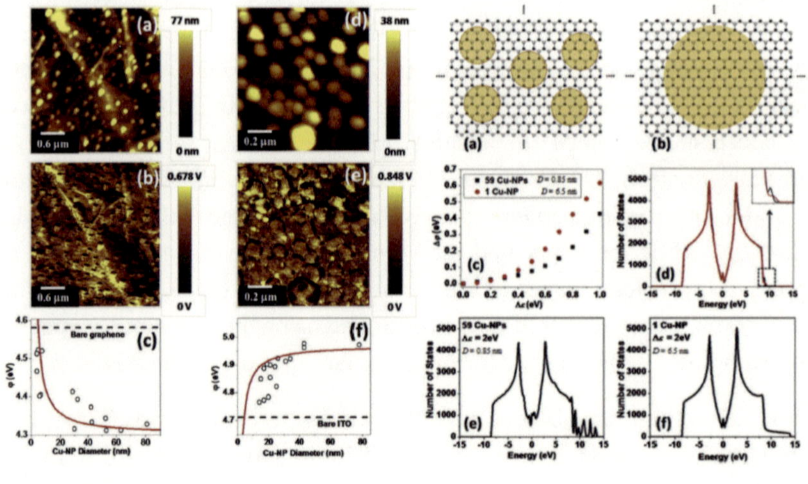

Figure 5. Left panel: (a) AFM and (b) SKPFM micrographs of graphene-based thin films decorated with Cu-NPs. (c) Plot of the work function vs. Cu-NP diameter. (d) AFM and (e) SKPFM micrographs of Cu-NPs on ITO. (f) Plot of the work function vs. Cu-NP diameter for Cu-NPs on ITO. **Right panel:** (a) 59 Cu-NPs of $D = 0.85$ nm diameter distributed randomly on a graphene lattice and (b) 1 Cu-NP of $D = 6.5$ nm. (c) Calculated shift in the work function. (d) The density of states for the two distributions indicated in panels a and b ($\Delta\varepsilon = 0.3$ eV). (e) and (f) show the density of states for the two distributions at $\Delta\varepsilon = 2$ eV. Reproduced with permission from Ref. [5].

The right panel of **Figure 5** describes the result of theoretical modeling of both the work function and the density of states (DOS) of graphene-based films decorated with either a random array of 59 small diameter Cu-NPs (0.85 nm) compared to a single large Cu-NP (6.5 nm). Graphene doping due to the Cu distribution can be well understood in the framework of a tight-binding model considering two different ionization energies, $\varepsilon_i = \varepsilon_0$ and $\varepsilon_j = \varepsilon_{mod}$, for two homogeneously distributed but non-equivalent carbon sites, *i* and *j*, interacting and non-interacting with Cu, respectively. The plot of the work-function shift as a function of the difference, $\Delta\varepsilon = \varepsilon_{mod} - \varepsilon_0$, of the local ionization energy of *i* and *j*-type carbon sites is reported in **Figure 5c** (right panel). For $\Delta\varepsilon > 0.3$ eV, the variations in work function change ($\Delta\varphi$) between the two systems are large, and shows that for stronger metal-carbon bonding the particle diameter plays a more significant role in shifting the work function away from the Dirac point of graphene, in addition to the major role played by the area fraction covered by metallic particles [5].

At small $\Delta\varepsilon$, the DOS is only marginally different for the two distributions of Cu shown in the right panel of Figures 5a and b, as demonstrated by DOS plot in the right panel d. An important feature of the DOSs at high-energy differences between the Dirac point of graphene and the work function of metallic nanoparticles, shown in right panels e and f ($\Delta\varepsilon = 2$ eV), shows the DOS is significantly affected by the particle size. If very small particles are randomly distributed on the graphene lattice, the DOS near the Fermi level is dramatically altered in its profile with the appearance of additional states, compared to pristine graphene (**Figure 5e** right panel). However, even at the largest values of $\Delta\varepsilon$, the DOS profile near the Dirac point is preserved for large metallic nanoparticles, as demonstrated in right panel f for $D = 6.5$ nm. This suggests that for particles of $D > 10$ nm, the metal particle diameter plays a negligible role in determining the work-function shifts of graphene domains. These theoretical findings corroborate the SKPFM experiments, indicating that the Cu-NP diameter has only a moderate effect on the work-function shift ($\Delta\varphi$) of graphene where larger particles do not have a greater effect compared to smaller ones.

Strongly interacting metals such as Ni and Co have also been investigated for modifying graphene using metal nanoparticles. Ni has been shown to strongly interact with graphene surfaces and can be used to etch vacancies in graphene sheets or be used as an electrode material in graphene-based devices [1]. Cobalt nanoparticles on single-layer graphene have also been used for band engineering to open a small band gap in single-layer graphene at room temperature, sufficient for use in semiconducting applications [12]. In this application, both the size and the distribution of Co nanoparticles affects the band gap formation, which was controlled by using a very slow metal deposition rate (0.02 Å/s). A band gap is induced when the nanoparticle size is such that local oxidation is induced in the graphene by strong interaction with Co *d*-states and upon exposure to air, resulting in localization of free electrons in graphene at oxide sites and modification of the graphene band structure [12].

2.4. Specific properties and applications of graphene thin films decorated with arrays of metal nanoparticles

Graphene thin films decorated with random distributions of metallic nanoparticles have been shown to enhance optical and electronic properties, and new devices such as evanescent waveguides could emerge through self-assembling of nano-metallic phases into superlattices. Typically, expensive techniques such as nanolithography and nano-contact printing, suitable only over small areas, are required to attain nano-patterned structures on graphene or other 2D materials. For use over large areas, alternative methods have been proposed for the spontaneous self-assembly of metal nanoparticles such as Cu on graphene-based films. In the left panel of **Figure 6a** is shown a schematic of Cu-NP self-assembly on graphene-based films from Ouyang *et al.* [8]. Large-area, graphene-based films were deposited from a graphene-RNA dispersion discussed previously, followed by deposition of a Cu metal layer using thermal evaporation at very low deposition rates (0.5 nm/min). The left panel of **Figure 6c** shows the graphene immediately after Cu deposition, and **Figure 6d** shows the nucleation of Cu-NPs into a self-assembled superlattice.

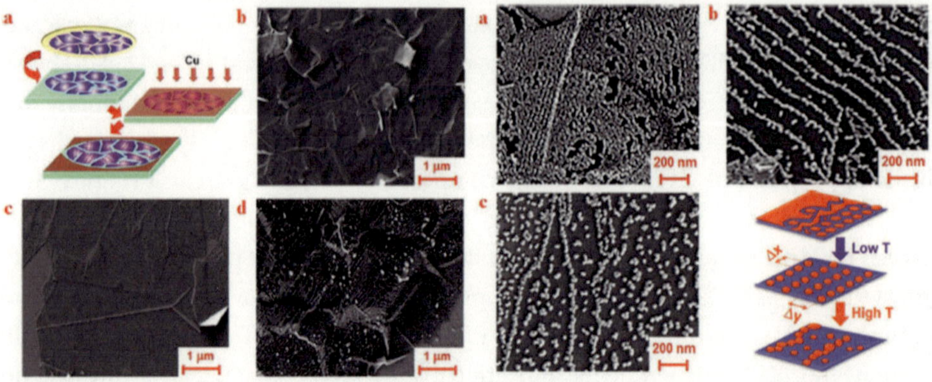

Figure 6. Left panel: (a) Fabrication of Cu-NPs superlattices: (i) deposition of a large-area graphene thin film; (ii) thermal evaporation of a 0.7 ± 0.3 nm-thick Cu layer, and (iii) thermal annealing of the system. (b) SEM of a bare graphene-based thin film on Si(100), (c) SEM of the same film after thermal evaporation of the Cu layer, and (d) SEM after thermal annealing. **Right panel:** SEM images of samples annealed for 8 h at increasing temperatures: (a) 200, (b) 360, and (c) 440°C. In the bottom right corner, a diagram shows the behavior of Cu-NP's nucleated at different temperatures. Reproduced with permission from Ref. [6].

The right panel of **Figure 6** shows the effect of different annealing temperatures that promote different conformations of Cu superlattices. The ideal annealing temperature was found to be 360°C, at which ordered superlattices of Cu-NPs with less than 20 nm diameter are created. After 8 h annealing at this temperature, Cu-NPs align in parallel lines along the armchair direction of the graphene lattice. The inter-particle spacing in this case is relatively low with a ~12 nm inter-particle spacing along the same line, while the spacing between lines is found to be one order of magnitude higher (~182 nm) [6]. The use of higher annealing temperatures' (**Figure 6**, right panel c) kinetic effects cause the Cu-NPs to become displaced from the superlattice structure observed at 360°C. At lower temperatures, shown in **Figure 6**, right panels a and b, Cu-NP nucleation occurs when the thin Cu layer is in a molten state since the melting temperature of metallic nanoparticles and ultrathin metallic films of only a few atomic layers is significantly lower than in the bulk metal.

The propagation of evanescent waves along Cu-NP superlattices confined in the proximity of the graphene surface and propagating along the direction parallel to the Cu-NPs lines was investigated using three-dimensional scanning near-field optical microscope (3D-SNOM). In these experiments, a 532 nm laser is launched into a single-mode optical fiber and incident on the superlattice samples as shown in **Figure 7d** and **e**. SNOM (X,Z) scans were used to probe the decay of near-field evanescent modes moving away from the sample surface along the *z*-direction. Vertically, evanescent fields only exist within a distance equal to approximately one wavelength of incident light. In the near field, Cu-NPs display strong Mie scattering to $\lambda = 532$ nm light, and therefore the interaction between light and the Cu-Np superlattices is dominated by scattering [6]. Light incident on the Cu-Np superlattice experiences Mie scattering along a preferential direction when the interline distance (~100 nm) is the same order of magnitude as the wavelength of incident light. Light absorption occurs in the transverse direction to the NP lines, resulting in low attenuation for longitudinal modes and confinement of an evanescent wave as shown in **Figure 7d** [6].

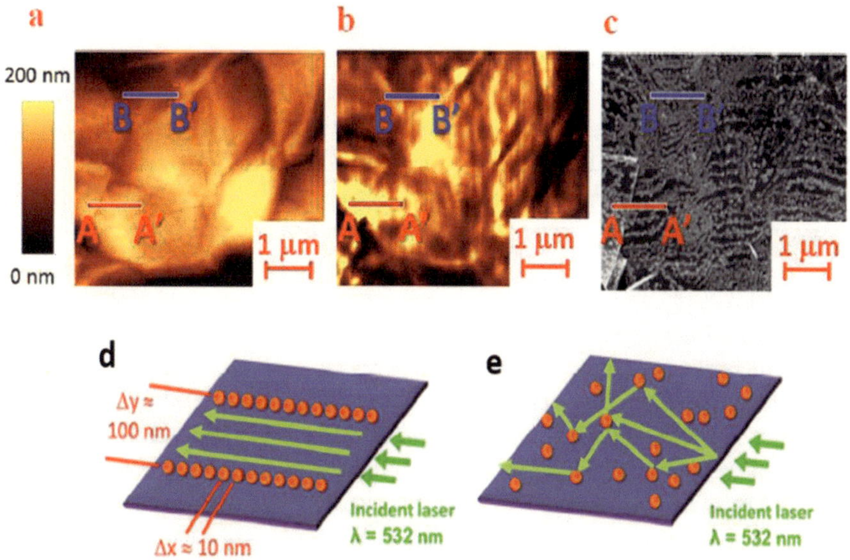

Figure 7. (a) AFM and (b) 3D-SNOM images obtained simultaneously from a Cu-NP superlattice annealed at 360°C for 8 h. (c) SEM micrograph obtained using lithographically patterned markers from the same sample region as in panels a and b. (d) Coherent scattering in superlattices and (e) incoherent scattering combined with light absorption in randomly distributed Cu-NPs. Reproduced with permission from Ref. [6].

Conversely, light will be isotropically scattered in all directions by randomly distributed Cu-NPs (**Figure 7e**), which causes interference between the far-field modes such that evanescent waves do not propagate. 3D-SNOM measurements in **Figure 7b** describe these two cases of evanescent waveguide behavior in Cu-NPs superlattices (A-A′ region) and isotropic Mie scattering where Cu-NPs are disordered (B-B′ region). The area of the AFM topography shown in **Figure 7a** is the same area defined in the SNOM image of **Figure 7b**, in which a brighter pixel represents a higher intensity of scattered light being collected at that point in close proximity to the sample at low z distances.

Figures 7b and **c** show a superlattice of Cu-NP's aligned parallel to the light propagation direction which is along the Y-axis. The SEM of **Figure 7c** shows that the Cu-NP distribution is ordered in the A-A′ region, and significantly less ordered in the region B-B′. From this result, the section A-A′ is expected to function as an evanescent waveguide, while B-B′ should exhibit a sequence of constructive and destructive interference patterns along the vertical axis, due to Mie scattering [6]. Along the A-A′ section, confinement of the electromagnetic radiation in the bright yellow zone in the proximity of the sample surface can be observed. A strong attenuation of light intensity at values of Z (not shown) within a few tens of nanometers from the sample surface was also observed, verifying the presence of evanescent modes in the A-A′ region.

3. Conclusions

Graphene and graphene-based films can be fabricated using a variety of methods including vacuum deposition techniques such as CVD, to solution-based techniques focusing on the exfoliation of graphite in aqueous solutions. CVD techniques are typically used to produce pristine grains of single- and few-layer graphene grown on Ni and Cu substrates. Graphene of this type is high quality, but can be costly to produce and requires an additional process often based on a sacrificial polymer layer to transfer the graphene to other substrates for device applications. Although the as-prepared graphene samples are of high quality, defects and contamination can be introduced during the transfer process, lowering the overall quality of the final product. Graphene-based materials produced from chemical exfoliation of graphite can overcome some issues with vacuum techniques where chemical exfoliation does not require vacuum facilities and is suitable to scaling up to large quantities. Graphene and graphene oxide dispersions in water obtained by oxidation or the use of a surfactant can be used to produce graphene-based films using simple deposition techniques such as vacuum filtration or spin coating. Although more suitable for large-scale production, chemical exfoliation methods may be less reproducible as the quality of the final graphene-based film depends strongly on the specific fabrication parameters.

It has been shown that films produced from interlocking and overlapping graphene platelets have optical, electrical, and thermal properties that diverge from single-layer graphene. These films may therefore be more desirable for various electronic and device applications, particularly if they are further modified with metal nanoparticles. In this application, the deposition and nucleation conditions of metal NPs determine the properties of the resulting graphene-NP films. The nucleation of metal NPs and their size can be controlled using slow metal deposition rates combined with long annealing times at relatively low temperatures in an inert atmosphere to avoid metal oxidation. These techniques can be used to obtain self-assembled arrays of NPs on graphene-based films that can be used in new types of nano-devices such as evanescent waveguides. It is clear that metal NP modified graphene materials are well positioned as new materials with enhanced properties to be used in the next generation of electronic devices.

Acknowledgements

PB acknowledges a MITACS Accelerate Postdoctoral Fellowship. AAS acknowledges an Ontario MRI Queen Elizabeth II Graduate Scholarship. JP acknowledges a Nanofabrication Facility Graduate Student Award from the University of Western Ontario. GF acknowledges a Canada Research Chair in carbon-based nanomaterials. The authors would like to thank Dr. M.S. Ahmed, Dr. R.J. Bauld, Dr. F. Sharifi, and Mr. A. Venter for some of the experiments that led to the results discussed here, and for fruitful discussions. Funding from the Canada Foundation for Innovation (grant no. 212442) and the Discovery Grant program of the Natural Sciences and Engineering Research Council of Canada (RGPIN-2015-06004) are also gratefully acknowledged.

Author details

Paul Bazylewski[1], Arash Akbari-Sharbaf[1], Sabastine Ezugwu[1], Tianhao Ouyang[1], Jaewoo Park[1] and Giovanni Fanchini[1,2*]

*Address all correspondence to: gfanchin@uwo.ca

1 Department of Physics and Astronomy, University of Western Ontario, London, Canada

2 Department of Chemistry, University of Western Ontario, London, Canada

References

[1] Zhang Y, Zhang L, Zhou C. Review of chemical vapor deposition of graphene and related applications. Accounts of Chemical Research. 2013;46:2329–2339. DOI: 10.1021/ar300203n

[2] Park S, Ruoff RS. Chemical methods for the production of graphenes. Nature Nanotechnology. 2009;5:309–309. DOI: 10.1038/nnano.2009.58

[3] Zhu Y, Murali S, Cai W, Li X, Suk JW, Potts JR, Ruoff RS. Graphene and graphene oxide: synthesis, properties, and applications. Advanced Materials. 2010;22:3906–3924. DOI: 10.1002/adma.201001068

[4] Sharifi F, Bauld R, Ahmed MS, Fanchini G. Transparent and conducting graphene-RNA-based nanocomposites. Small. 2012;8:699–706. DOI: 10.1002/smll.201101537

[5] Akbari-Sharbaf A, Ezugwu S, Ahmed MS, Cottam MG, Fanchini G. Doping graphene thin films with metallic nanoparticles: experiment and theory. Carbon. 2015;95:199–207. DOI:10.1016/j.carbon.2015.08.021

[6] Ouyang T, Akbari-Sharbaf A, Park J, Bauld R, Cottam MG, Fanchini G. Self-assembled metallic nanoparticle superlattices on large-area graphene thin films: growth and evanescent waveguiding properties. RSC Advances. 2015;5:98814–98821. DOI: 10.1039/c5ra22052a

[7] Ahmed MS, Ezugwu S, Divigalpitiya R, Fanchini G. Relationship between electrical and thermal conductivity in graphene-based transparent and conducting thin films. Carbon. 2013;61:595–601. DOI:10.1016/j.carbon.2013.05.041

[8] Bauld R, Ahmed MS, Fanchini G. Optoelectronic and thermal properties of solution-based graphene thin films. Physica Status Solidi C. 2012;9:2374–2379. DOI: 10.1002/pssc.201200292

[9] Bauld R, Sharifi F, Fanchini G. Solution processed graphene thin films and their applications in organic solar cells. International Journal of Modern Physics B, 2012;26:1242004. DOI: 10.1142/S0217979212420040

[10] Venter A, Hesari M, Ahmed MS, Bauld R, Workentin MS, Fanchini G. Facile nucleation of gold nanoparticles on graphene-based thin films from Au144 molecular precursors. Nanotechnology. 2014;25:135601. DOI:10.1088/0957-4484/25/13/135601

[11] Bauld R, Hesari M, Workentin MS, Fanchini G. Thermal stability of Au_{25} molecular precursors and nucleation of gold nanoparticles in thermosetting polyimide thin films. Applied Physics Letters. 2013;101:243114. http://dx.doi.org/10.1063/1.4770515

[12] Bazylewski PF, Nguyen VL, Bauer RPC, Hunt AH, McDermott EJG, Leedahl BD, Kukharenko AI, Cholakh SO, Kurmaev EZ, Blaha P, Moewes A, Lee YH, Chang GS. Selective area band engineering of graphene using cobalt-mediated oxidation. Scientific Reports. 2015;5:15380. doi:10.1038/srep15380

Chapter 2

Possible Role of Microcrystallinity on Surface Properties of Titanium Surfaces for Biomedical Application

Federico Mussano, Tullio Genova,
Salvatore Guastella, Maria Giulia Faga and
Stefano Carossa

Additional information is available at the end of the chapter

http://dx.doi.org/10.5772/62914

Abstract

Dental implantology has grown tremendously, since the introduction of titanium. To enhance osseointegration, roughening techniques such as grit blasting, chemical etch, electrochemical anodization have been used with good results. An oxide layer mainly composed of TiO_2 covers the surface of dental implants ensuring excellent corrosion resistance and chemical stability. Despite its biological role in achieving bone interlock, surprisingly, little is known about the structure of TiO_2, which may be either amorphous or crystalline. Furthermore, at least two crystalline polymorph phases can be found at the bone–implant interface: anatase (tetragonal) and rutile (tetragonal). Therefore, besides the recognized importance of surface topography, energy, and charge, a more refined knowledge of surface chemistry is advisable when studying the bone–implant interface. Recently, sophisticated analysis techniques have been applied to dental implants such as Raman spectroscopy and X-ray diffraction to obtain structural-crystallographic characterization.

This book chapter reviews the scientific literature with the scope of assessing what is known about the surface micro-/nanotopography and the crystallographic microstructure of titanium dental implants. Also, the correlation between these surface features and the biological outcomes in vitro and in vivo is a primary object of the manuscript. An electronic search was made in the databases of MEDLINE (through MeSH) and SCOPUS, extended to September 30th 2015, with no linguistic restrictions.

Based on the results of the most recent studies, the surface of titanium dental implants may be constituted of anatase, rutile, and amorphous phases. Anatase seems more present in arc-oxidized implants, alone or with rutile, according to the oxidation conditions (voltage, electrolyte etc.). Rutile and amorphous phases are more frequently found in machined, double-etched, sandblasted, sandblasted acid-etched implants. Particular interest is raised by the possible presence of brookite, which was found on a commercially available sandblasted acid-etched implant. Taking into consideration the variations in the biological activity of these polymorphs, identification of the TiO_2 phases found in the surface layers of implants should be regarded as fundamental not only by researchers but also by manufacturers.

Keywords: Raman spectroscopy, Dental implants, Nanotopography, Surface micro-crystallography, Surface properties

---

1. Osseointegration: an overview of clinically used surfaces

Since Swedish orthopedic surgeon and researcher Per-Ingvar Brånemark discovered the particular connection titanium was capable to develop within bone [1], the concept of osseointegration has been developed as a stable and direct interlock between bone and implant [2, 3]. Currently, commercially pure titanium (Grade 4 titanium) and Ti–6Al–4V alloy (Grade 5 titanium) have become the preferred material in implant dentistry, although ceramic materials with the use of zirconium dioxide and innovative metallic alloys are also attracting growing interest in dentistry [4]. Indeed, the number of dental implant brands on the market increased remarkably during the last three decades from 45 systems in 1988 [5], to 225 systems in 2002 [6], reaching an estimate of 1600 systems nowadays.

In such a competitive field, among all the possible approaches experimented in order to improve the properties of titanium implant surfaces, the main route adopted by the research and industry to enhance osseointegration has successfully entailed roughening techniques [7, 8]. Briefly, the different essential types of modification available on the market can be achieved by applying physical or chemical agents on the implant surface, as follows:

a. blasting (sand, glass or ceramic microspheres accelerated toward the surface);

b. wet etching (exposition to acid or alkali chemicals);

c. anodization;

d. plasma spray;

e. exposition to laser radiation;

f. exposition to electron beams.

Other treatments will be briefly outlined including exposition to cold plasmas and inorganic coatings.

Abrasive blasting (also called sandblasting or grit blasting) is a very common type of surface modification, thanks to the simplicity, low cost, and easy application. Microspheres of diameter in the range 10–540 μm are typically accelerated toward the surface to be treated, using a compressed air or nitrogen blow. Corundum (Al_2O_3) [9, 10], silicon carbide (SiC) [11], titania (TiO_2) [12], hydroxyapatite (HA) [13], zirconia (ZrO_2) [14], silica (SiO_2) [15], and aluminum powders [15] are the most used grit materials. Increasing roughness is the main effect sandblasting obtains on the morphology of the treated surface. Several parameters contribute to the roughening process, including: the material type, the sphere dimension, the treatment duration, and the energy and angle at the moment of the impact on the surface. The roughness of dental implants normally spans from Ra = 0.3 μm to Ra = 3 μm [15, 16], while polished Ti surfaces assume Ra values lower than 0.1 μm [15, 16]. A side effect of the sandblasting process is, however, the contamination of the surface resulting from the material released by the microspheres during their interaction with the surface. Recently, it has been pointed out [15] that the different types of grit materials and the microsphere dimensions can lead to different amounts of surface contamination. In particular, alumina blasting with microspheres of 54 μm diameter was found to effectively remove Si contamination from the machined titanium surface, but it was also responsible for the Al contamination as high as ~15%.

Acid treatment is often used to remove contamination and obtain clean and uniform surface finishes. A combination of acids such as HCl, H_2SO_4, HNO_3, and HF is frequently used to pretreat titanium. A solution composed of 10–30 vol% of HNO_3 and 1–3 vol% of HF in distilled water has been recommended to be a standard solution for acid pretreatment. To reduce the possible incorporation of hydrogen in titanium and thus the embrittlement of the surface layer, a ratio of nitric acid to hydrofluoric acid of 10–1 is suggested [17]. Acid etching generally leads to a thin surface oxide layer (<10 nm). Although the oxide is predominantly TiO_2, residues from the etching solution are frequently observed, especially chemicals containing fluorine.

Of great interest is the dual thermo-etching process first proposed by Beaty that has become the paradigm for the dual acid-etched surfaces [18]. Titanium surface is immersed in 15% HF solution and then etched in a mixture of H_2SO_4/HCl (6:1) and heated at 60–80°C for 3–10 min. The main effect of the acid-etching processes is to modify the implant morphology by producing micropits of a few microns diameter on titanium surfaces [16, 19] (**Figure 1A**).

Acid etching is also commonly applied after sandblasting. The complete process, usually referred to as sandblasting and large grit acid etching (SLA) [20], is often considered the reference surface treatment in dental implantology [12, 16, 1920]. This process and its derivatives involve the use of alumina microspheres of 200–540 μm diameter, followed by the etching with a mixture of HCl and H_2SO_4 [16] (**Figure 1B**). The SLA surface treatment combines the

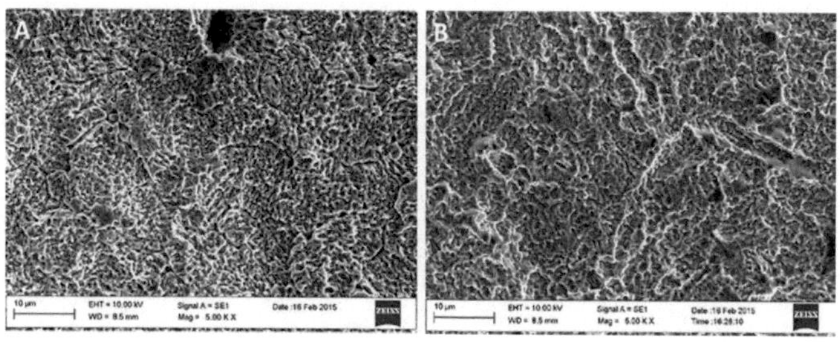

Figure 1. SEM images of a dual acid-etched surface treatment commercially known as DM (A) and a sandblasted acid-etched surface treatment commercially known as SL (B). The samples were provided courtesy of Titanmed s.r.l. (Milan, Italy).

macroroughness generated by the sandblasting process with the microroughness achieved through the acid-etching process [21].

Employed together, alkali and heat treatment [22] enable the formation of a biologically active bone-like apatite layer on the surface of titanium [23]. Due to the strong tendency of titanium to oxidize, the heat treatment is performed at a pressure of 10^{-4}–10^{-5} Torr. Crystalline sodium titanate (when using NaOH as a base) as well as rutile and anatase precipitates after thermal treatment. The whole process generates a surface capable of promoting the HA precipitation in simulated body fluid following Kokubo's test [ISO 23317:2014(E)].

A native oxide layer grows slowly and spontaneously on titanium kept in air, with an estimated rate of 3–6 nm during a 400-day period [24]. To substitute this thin layer with a thick porous layer of titanium oxide, anodization is widely used. This process consists in either a potentiostatic or a galvanostatic electrochemical oxidation, usually carried out in strong acids, such as HNO_3, H_2SO_4, H_3PO_4, and HF [19, 25]. To some extent, it is also possible to choose the phase of the titanium oxide layer among its amorphous, brookite, rutile, and anatase forms [25].

Titanium plasma spraying (TPS) consists in projecting titanium powders onto the implant surface by means of plasma torch at high temperature. Thus, the titanium particles condense and fuse together, forming a film about 30–50 μm thick [4]. The resulting coating has an average roughness of Sa 4 μm [26]. This three-dimensional topography was reported to increase the tensile strength at the bone–implant interface in vivo [20].

In an endless endeavor to improve the properties of Ti surfaces [27–30], laser treatments have also been proposed. As a result of the heating generated by the absorption of the high-density radiation, the main effect of laser radiation on metals, such as Ti, is to produce a localized melting of the material. The melting process involves only a very thin metal layer under the surface, which is quickly recrystallized after the radiation beam is moved to another portion of the surface, while a titanium oxide layer is formed because of the interaction between solidifying metal and air [27]. Although several types of lasers suit for the modification of metals and oxides, including ruby, like Nd–YAG, argon ion, CO_2 and excimer lasers [29], Nd–YAG appears to be the most diffused one for titanium and its alloys in dentistry [27–30]. Laser-treated Ti is usually rougher than machined Ti surfaces, with typical Ra values ranging from 0.5 to 2 μm [29, 30].

Electron beams have been introduced [31, 32] and used mainly as a pretreatment for the deposition of CaP coatings on titanium [31]. The process was found to reduce the roughness while improving the nanohardness of the material [32] and permitting the deposition of smoother CaP layers [31].

As plasma treatments could prove advantageous compared to wet techniques, such as acid etching, owing to the absence of chemical residuals on the surface, the avoidance of chemical waste, and the reduced safety concerns during manufacturing [33], their application has greatly increased recently.

Depending on the pressure conditions at which they are carried out, plasma treatments can be subdivided into vacuum plasma treatments (reduced pressure plasma treatments) and atmospheric pressure plasma (APP) treatments. APP treatments are simple and user friendly, however, when dealing with industrial application, reduced pressure plasma displays some advantages. At low pressure, a lower power is required to activate a plasma discharge and, even more importantly, the process performed in vacuum ensures a controlled environment less prone to external contaminations. Although plasma processes have mostly been applied for cleaning and sterilizing dental implants, owing to their capacity to strongly affect the surface energy, they have also been tested for the acceleration of osseointegration [33–36] and the application of antibacterial features to implants [33, 34, 37]. To this aim, argon and oxygen were preferably selected [33–38]. Speaking of plasma treatments, plasma immersion ion implantation (PIII) techniques are also noteworthy as a promising future research avenue in intrabony biomaterials. Here, plasma is used as a source of ions, which are accelerated toward the treated surface and there implanted [38]. Very recently, the incorporation of specific chemical elements such as fluorine (F) [39], calcium (Ca) [40], and zinc [41] was described to confer suitable biological properties.

For the sake of completeness, it is convenient to briefly outline some additive surface modifications, in spite of their limited human use. Calcium phosphate (CaP)-based alloys [42, 43] including HA [$Ca_{10}(PO_4)_6(OH)_2$] [42] and calcium phosphate cements (CPC) [43] result among the most studied coating materials for the enhancement of osseointegration. Several methods have been tested for the deposition of CaP coating on Ti implants, including plasma spray, sputtering, sol–gel deposition, and electrophoretic deposition processes, but plasma spray is considered the most successful so far [33]. Plasma-sprayed coatings can be deposited with a thickness ranging from a few micrometers to a few millimeters, which are characterized by their own roughness and show low density and high porosity [44]. Within the body fluids, these materials lead to the formation of HA nanocrystals. Calcium plays a relevant role in binding biologically active proteins as in its ionized form it adsorbs to the TiO_2 surface and further to macromolecules with high affinity for Ca^{2+} [45, 46].

Plasma-sprayed HA coatings are usually composed of large crystalline HA particles embedded into a highly soluble amorphous calcium phosphate phase. Numerous clinical studies were reported for HA-coated implants [47–49]. Unfortunately, plasma-sprayed HA-coated dental implants have been associated with clinical problems [50–54], due to the possible delamination of the coating from the bulk underneath. This break at the implant-coating interface obviously implies the implant failure despite the fact that the coating is well attached to the bone tissue [50–53]. Coating delamination has been described when the efficacy of plasma spraying was not optimal owing to the size of the dental implants [50]. Loosening of the coating has been reported, especially when the implants have been inserted into dense bone. Inflammatory reaction caused most of long-term failures. Tsui et al. [55, 56] associated the presence of metastable and amorphous phases in the HA coating during the plasma-spraying process to the low crystallinity of HA coating and to the deriving poor mechanical strength [57]. Despite their negative reputation in dental practice, a meta-analytic review could not detect significantly inferior long-term survival rates of plasma-sprayed HA-coated dental implants compared to other dental implants [51].

2. Key surface features

Accurate surface characterization is a fundamental topic in material science. Several relevant surface parameters can be characterized easily using standard analytical methods, such as contact or optical profilometry, electron microscopy and contact angle determination, independently of the production process. This permits to classify the surface of a given implant based on two key characteristics:

i. topography at the microscale (roughness) and nanoscale;

ii. wettability

2.1. Topography

At the microscale, the topography of an implant surface is supposed to increase the contact surface and thus the biomechanical interlocking between bone and implant [58]. However, as bone is continuously remodeled [59], the functional osseointegrated area is lower than the theoretical surface developed area [60]. The effects of the various microtopography patterns on bone apposition are still unclear and require more investigations. The quantitative description of surface topography is usually based on roughness, which can be determined either as a profile (2d) or evaluating the whole area (3d). In the former case Ra, Rz, and Rms are the key parameters, while in latter case, it occurs to mention Sa, Sds, and Sdr%. Height deviation amplitude (Sa) is conveniently used for classifying osseointegrated implants into four categories: smooth 0–0.4 μm, minimal 0.5–1 μm, moderate 1–2 μm, and maximal >2 μm [58, 61]. As for spatial density, surfaces are either rugged (Sdr% > 100%) or flattened out (Sdr% < 100%), while the morphology of the microstructures may be described as rough, patterned, or particled, with respect to the number of dimensions. Specifically, following Dohan Ehrenfest et al., "microrough surfaces have one micrometric dimension (the peak heights). Micropatterns have two micrometric dimensions (dimensions of the repetitive pattern), such as the micropores created by anodization (…). Microparticles have three micrometric dimensions."

At the nanoscale, a more textured surface topography is known to increase the surface energy. The higher the surface energy the greater becomes the wettability. To an increased, wettability is due to the improved adsorption of fibrin and matrix proteins on the surface, which, in turn, favors cell attachment, tissue healing, and eventually the osseointegration process. Nanotopography might also directly influence cell behavior through the influence nanopatterning has on the cytoskeleton [62–66]. Even though all surfaces have their own nanotopography, by definition, not all of them possess significant nanostructures. A nanostructure is conventionally defined as an object of size comprised between 1 and 100 nm. Dealing with nanostructures, it may be helpful to specify the number of nanoscale dimensions. One dimension at the nanoscale implies the concept of nanoroughness [67], while nanopatterns are endowed with two nanoscale dimensions, like the nanotubes produced by anodization [68, 69], or the chemically produced nanopatterned surfaces [64, 70] (**Figure 2**).

http://dx.doi.org/10.5772/62914

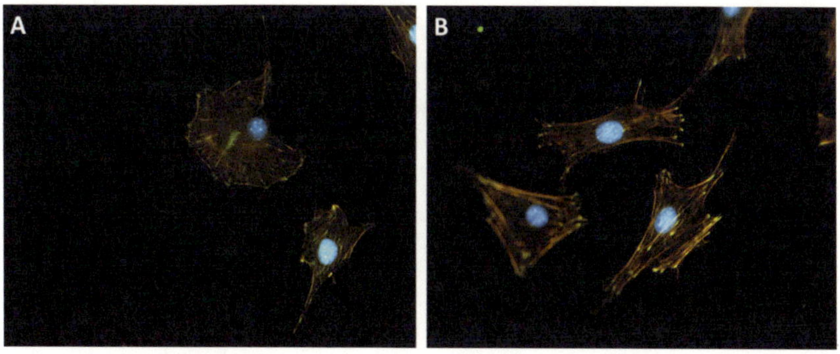

Figure 2. The picture depicts morphologically the cytoskeleton arrangement of murine osteoblasts (MC3T3-E1) grown, respectively, on smooth (A) and nanostructured titania surfaces (B). Cells were stained with Rodamine–phalloidin (red), anti-paxillin antibody (green), and DAPI (blue). The effect of the surface pattern on the cells is clearly appreciable from the number of focal adhesions as visualized by marking paxillin in green.

The presence of three nanoscale dimensions is typical of the nanoparticles. If nanostructures are not clearly detectable (no patterns, no particles, insignificant texture) or not homogeneous and repetitive, the surface should be considered as nanosmooth. Repetitiveness and homogeneity are indeed important yet difficult to define—morphological parameters that may be deemed qualitative.

2.2. Wettability

The wetting features of a solid material are usually determined through the sessile drop technique. Briefly, a drop of a given liquid is placed on the surface sample and the angle between the tangent of this drop at the three-phase boundary and the solid surface is measured. Thus, the contact angle CA expressing the surface wetting is quantified according to the liquid employed. For instance, if water is used, the CA will characterize the hydrophilicity of the surface. In principle, the CA can assume values from 0° to 180°, in case of complete spreading or beading of the drop, respectively. Water CAs lower than 90° ascribe surfaces a hydrophilic feature, while water CAs above 90° designate surfaces as hydrophobic. As the "the drop rests on an ideal homogeneous and flat surface in thermodynamic equilibrium, the drop shape with the characteristic ideal CA θ is formed as a result of the liquid/vapor γlv, solid/liquid γsl, and solid/vapor γsv interfacial tensions, according to Young's equation". (...) Surface tension is caused by the asymmetry of the cohesive forces of molecules at a surface compared to molecules in the bulk where each molecule has surrounding partners resulting in a net force of zero. Correspondingly, the surface energy is minimized in the bulk, whereas at the surface, the energy is increased due to the missing surrounding molecules. Therefore, to reduce surface energy, the surface area has to be minimized, thus resulting in phenomena like spherical water drops or the spreading of aqueous liquids on higher energetic surfaces." [71]

High energetic solid surfaces enhance wetting, which has been associated with improved implant success [72] (**Figure 3**).

A

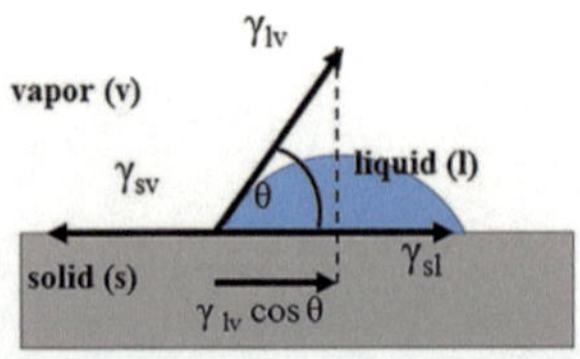

B

Optical Representation	Macroscopic Result	CA Relationship	Energetic Relationship
	Non-wettability	$\theta = 180°$	$\gamma_{sl} - \gamma_{sv} > \gamma_{lv}$
	Low wettability	$90° \leq \theta \leq 180°$	$\gamma_{sv} - \gamma_{sl} < 0$
	High wettability	$0° \leq \theta \leq 90°$	$\gamma_{sv} - \gamma_{sl} > 0$
	Complete wetting	$\theta = 0°$	$\gamma_{sv} - \gamma_{sl} > \gamma_{lv}$

Figure 3. (A) Schematic diagram depicting contact angle CA as measured by sessile drop technique, along with the graphical derivation of Youngs equation. (B) Relations between wetting tension and the wetting of a solid surface. Figure concept has been taken from Rupp et al. [71].

3. The possible role of microcrystallinity state of the titanium surface

The outstanding chemical inertness, repassivation ability, corrosion resistance, and ultimately biocompatibility of titanium result from an oxide layer that is usually only a few nanometers thick. As titanium exists in many different stable oxidation states and oxygen is highly soluble in titanium, titanium oxide is known to have varying stoichiometries. Among the common compounds, there are Ti_3O, Ti_2O, Ti_3O_2, TiO, Ti_2O_3, Ti_3O_5, and TiO_2 [73]; however, the most stable titanium oxide is TiO_2, also known as titania. TiO_2 is thermodynamically very stable and the Gibbs free energy of formation is highly negative for a variety of oxidation media, such as water or oxygen containing organic molecules.

Although the fundamental biological role of titania in osseointegration has attracted a lot of interest, there is limited knowledge regarding its structure, especially on commercially available products. TiO_2 exists in three crystalline polymorph phases: rutile (tetragonal), anatase (tetragonal), and brookite (orthorhombic), but only rutile and anatase phases are practically important. Brookite is the largest phase, with eight titania groups per crystal unit cell, anatase possesses four groups per unit cell, and finally, rutile has two groups per unit cell. Rutile is the most diffused and stable isoform. In all phases, a six-coordinated Ti participates in unit cells [74]. Titania may be found on implant surfaces in either the amorphous or crystal phase with heterogeneous results [75], as a consequence of the surface treatment the implants underwent [76, 77]. X-ray diffraction (XDR) is the technique of choice whenever the crystalline structure is to be investigated, for instance, in terms of main crystal orientation, grain size, crystallinity, and strain [78]. X-ray photoelectron spectroscopy (XPS) is instead used to determine the quantitative mean atomic composition of wide and thin surface areas (typically 300 mm in diameter, 5–7 nm depth). When XPS is applied to pure titanium samples exposed to the atmosphere at room temperature after milling, beside the stable titania film, hydroxide, and chemisorbed water bond with Ti cations are detected on the surface. In addition, some organic species like hydrocarbons adsorb and alkoxides or carboxylates of titanium also exist on the outmost surface layer. Currently, microcrystallinity has almost never been assessed in commercially available surfaces [79, 80].

During implant manufacturing, anatase, rutile, or amorphous TiO_2 are produced depending on the conditions. Upon heating, amorphous titania converts to anatase (<400°C) and then to rutile (600–1000°C) [81, 82]. The two crystalline phases, and especially anatase, have been studied as regards photocatalysis and photon–electron transfer [83], hydrophilicity [84], and biological decontamination capacity [85]. Notwithstanding its increased biological activity [79, 86], anatase has been claimed to be more prone to ionic dissolution in than rutile [87]. On the other hand, rutile renders surfaces hydrophobic, whereas anatase improves wetting [85], which may be beneficial for the healing process at early stages.

Recently, these properties have attracted growing interest, as they may provide a synergistic effect to the wide range of the surface treatments used. As mentioned above, the information available on the TiO_2 phases formed on the implant surfaces present in the market is surprisingly limited. The rapid growth of the oxide layer during manufacturing is thought to lead to an amorphous phase on implant surfaces [88]. Despite the well-documented interaction of amorphous TiO_2 layer with bone, HA cannot readily grow on such a surface, in simulated bodily fluid, which may be due to the arrangement of the oxygen portions. In rutile and anatase, however, oxygen groups match better the hydroxyl groups of HA, resulting in deposition of biomimetic apatite, thus possibly facilitating osseointegration [88]. As these phases require additional treatments to be grown from native amorphous TiO_2, Gaintantzopoulou et al. [89] hypothesized that the various surface treatments performed on titanium implants to enhance osseointegration were aimed at creating anatase and rutile crystalline domains. Briefly, they found that anatase is more pronounced in arc-oxidized implants, alone or with rutile, dependent on the oxidation conditions. Rutile and/or amorphous phases are more common in machined, double-etched, sandblasted, sandblasted acid-etched.

4. Conclusions

Distinct minorities of the implant manufacturers have undertaken basic, animal and human research when designing new or altering the components of existing implant systems. Consequently, many currently commercially available dental implants have insufficient, questionable, or simply totally lacking scientific justification of the product designs and material compositions. Potential alterations of the implants include surface chemical and biochemical properties, corrosion characteristics and wear debris release, surface energy and wettability as well as topography on micrometer and nanometer scales. Considering the possible role in their biological activity, the identification of the titania phases found in the surface layers of implants should be deemed unavoidable by the manufacturers and the scientific community.

Author details

Federico Mussano[1*], Tullio Genova[1,2], Salvatore Guastella[4], Maria Giulia Faga[3] and Stefano Carossa[1]

*Address all correspondence to: federico.mussano@unito.it

1 Department of Surgical Sciences, CIR Dental School, UNITO, Turin, Italy

2 Department of Life Sciences and Systems Biology, UNITO, Turin, Italy

3 IMAMOTER-National Council of Research, Turin, Italy

4 D.SAT Polytechnic University of Turin, Turin, Italy

References

[1] Albrektsson T, Sennerby L. State of the art in oral implants. J Clin Periodontol [Internet]. 1991 [cited January 31 2016];18(6):474–81. Available from: http://www.ncbi.nlm.nih.gov/entrez/query.fcgi?cmd=Retrieve&db=PubMed&dopt=Citation&list_uids=1890231

[2] Zarb GA, Smith DC, Levant HC, Graham BS, Staatsexamen WZ. The effects of cemented and uncemented endosseous implants. J Prosthet Dent [Internet]. 1979 [cited February 8 2016];42(2):202–10. Available from: http://www.ncbi.nlm.nih.gov/pubmed/287799

[3] William R. Dental Implants: Benefit and Risk. Proceedings of the NIH-Harvard Consensus Development Conference. 1978. pp. 1–351.

[4] Duraccio D, Mussano F, Faga M. G. Biomaterials for dental implants: current and future trends. J Mater Sci [Internet]. 2015;50(14):4779–812. Available from: http://dx.doi.org/10.1007/s10853-015-9056-3

[5] English CE. Cylindrical implants. Parts I, II, III. Calif Dent Assoc J. 1988;16:17–38.

[6] Jokstad A, Braegger U, Brunski JB, Carr AB, Naert I, Wennerberg A. Quality of dental implants. Int Dent J [Internet]. 2003 [cited January 3 2016];53(6 Suppl 2):409–43. Available from: http://www.ncbi.nlm.nih.gov/pubmed/14725421

[7] Cochran DL, Schenk RK, Lussi A, Higginbottom FL, Buser D. Bone response to unloaded and loaded titanium implants with a sandblasted and acid-etched surface: a histometric study in the canine mandible. J Biomed Mater Res [Internet]. 1998 [cited February 8 2016];40(1):1–11. Available from: http://www.ncbi.nlm.nih.gov/pubmed/9511093

[8] Wennerberg A, Hallgren C, Johansson C, Danelli S. A histomorphometric evaluation of screw-shaped implants each prepared with two surface roughnesses. Clin Oral Implants Res [Internet]. 1998 [cited February 8 2016];9(1):11–9. Available from: http://www.ncbi.nlm.nih.gov/pubmed/9590940

[9] Schweikl H, Müller R, Englert C, Hiller K-A, Kujat R, Nerlich M, et al. Proliferation of osteoblasts and fibroblasts on model surfaces of varying roughness and surface chemistry. J Mater Sci Mater Med [Internet]. 2007 [cited Feb 8 2016];18(10):1895–905. Available from: http://www.ncbi.nlm.nih.gov/pubmed/17546411

[10] Iwaya Y, Machigashira M, Kanbara K, Miyamoto M, Noguchi K, Izumi Y, et al. Surface properties and biocompatibility of acid-etched titanium. Dent Mater J [Internet]. 2008 [cited Feb 8 2016];27(3):415–21. Available from: http://www.ncbi.nlm.nih.gov/pubmed/18717170

[11] Yang G-L, He F-M, Yang X-F, Wang X-X, Zhao S-F. Bone responses to titanium implants surface-roughened by sandblasted and double etched treatments in a rabbit model. Oral Surg Oral Med Oral Pathol Oral Radiol Endod [Internet]. 2008 [citedx Feb 82016]; 106(4):516–24. Available from: http://www.ncbi.nlm.nih.gov/pubmed/18602288

[12] Kim H, Choi S-H, Ryu J-J, Koh S-Y, Park J-H, Lee I-S. The biocompatibility of SLA-treated titanium implants. Biomed Mater [Internet]. 2008 [cited 2016 Feb 8];3(2):025011. Available from: http://www.ncbi.nlm.nih.gov/pubmed/18458368

[13] Park J-W, Jang I-S, Suh J-Y. Bone response to endosseous titanium implants surface-modified by blasting and chemical treatment: a histomorphometric study in the rabbit femur. J Biomed Mater Res B Appl Biomater [Internet]. 2008 [cited Feb 8 2016];84(2): 400–7. Available from: http://www.ncbi.nlm.nih.gov/pubmed/17595031

[14] Bacchelli B, Giavaresi G, Franchi M, Martini D, De Pasquale V, Trirè A, et al. Influence of a zirconia sandblasting treated surface on peri-implant bone healing: an experimental study in sheep. Acta Biomater [Internet]. 2009 [cited Feb 1 2016];5(6):2246–57. Available from: http://www.ncbi.nlm.nih.gov/pubmed/19233751

[15] Guo CY, Matinlinna JP, Tsoi JKH, Hong Tang AT. Residual contaminations of silicon-based glass, alumina and aluminum grits on a titanium surface after sandblasting. Silicon [Internet]. 2015 [cited Feb 8 2016]; Available from: http://link.springer.com/10.1007/s12633-015-9287-6

[16] Rupp F, Scheideler L, Olshanska N, de Wild M, Wieland M, Geis-Gerstorfer J. Enhancing surface free energy and hydrophilicity through chemical modification of microstructured titanium implant surfaces. J Biomed Mater Res A [Internet]. 2006 [cited Feb 8 2016];76(2):323–34. Available from: http://www.ncbi.nlm.nih.gov/pubmed/16270344

[17] American Society for Testing and Materials, (1997), ASTM standard B600, Annual Book of ASTM standard, American Society for Testing and Materials, Philadelphia, PA., Vol. 2.04, p. 6.

[18] Baier, R.E; A.E. Meyer "Implant Surface Preparation," International Journal of Oral & Maxillofacial Implants, vol. 3, 9–20, 1988.

[19] Le Guéhennec L, Soueidan A, Layrolle P, Amouriq Y. Surface treatments of titanium dental implants for rapid osseointegration. Dent Mater [Internet]. 2007 [cited 2014 Nov 12];23(7):844–54. Available from: http://www.ncbi.nlm.nih.gov/pubmed/16904738

[20] Buser D, Schenk RK, Steinemann S, Fiorellini JP, Fox CH, Stich H. Influence of surface characteristics on bone integration of titanium implants. A histomorphometric study in miniature pigs. J Biomed Mater Res [Internet]. 1991 [cited 2015 Dec 30];25(7):889–902. Available from: http://www.ncbi.nlm.nih.gov/pubmed/1918105

[21] Szmukler-Moncler S, Perrin D, Ahossi V, Magnin G, Bernard JP. Biological properties of acid etched titanium implants: effect of sandblasting on bone anchorage. J Biomed Mater Res B Appl Biomater [Internet]. 2004 [cited 2016 Feb 8];68(2):149–59. Available from: http://www.ncbi.nlm.nih.gov/pubmed/14737762

[22] Kim HM, Miyaji F, Kokubo T, Nakamura T. Preparation of bioactive Ti and its alloys via simple chemical surface treatment. J Biomed Mater Res [Internet]. 1996 [cited 2016 Feb 8];32(3):409–17. Available from: http://www.ncbi.nlm.nih.gov/pubmed/8897146

[23] Kitsugi T, Yamamuro T, Nakamura T, Kakutani Y, Hayashi T, Ito S, et al. Aging test and dynamic fatigue test of apatite-wollastonite-containing glass ceramics and dense hydroxyapatite. J Biomed Mater Res [Internet]. 1987 [cited 2016 Feb 8];21(4):467–84. Available from: http://www.ncbi.nlm.nih.gov/pubmed/3034911

[24] Gurzawska K, Svava R, Yihua Y, Haugshøj KB, Dirscherl K, Levery SB, et al. Osteoblastic response to pectin nanocoating on titanium surfaces. Mater Sci Eng C [Internet]. 2014 [cited 2016 Feb 8];43:117–25. Available from: http://www.ncbi.nlm.nih.gov/pubmed/25175196

[25] Roach MD, Williamson RS, Blakely IP, Didier LM. Tuning anatase and rutile phase ratios and nanoscale surface features by anodization processing onto titanium substrate surfaces. Mater Sci Eng C Mater Biol Appl [Internet]. 2016 [cited 2016 Jan 21];58:213–23. Available from: http://www.ncbi.nlm.nih.gov/pubmed/26478305

[26] Canullo L, Genova T, Tallarico M, Gautier G, Mussano F, Botticelli D. Plasma of argon affects the earliest biological response of different implant surfaces: an in vitro com-

parative study. J Dent Res [Internet]. 2016 [cited 2016 Feb 8]; Available from: http://jdr.sagepub.com/cgi/doi/10.1177/0022034516629119

[27] Tian YS, Chen CZ, Li ST, Huo QH. Research progress on laser surface modification of titanium alloys. Appl Surf Sci [Internet]. 2005 [cited 2016 Feb 8];242(1–2):177–84. Available from: http://www.sciencedirect.com/science/article/pii/S0169433204012759

[28] Braga FJC, Marques RFC, Filho E de A, Guastaldi AC. Surface modification of Ti dental implants by Nd:YVO4 laser irradiation. Appl Surf Sci [Internet]. 2007 [cited 2016 Feb 8];253(23):9203–8. Available from: http://www.sciencedirect.com/science/article/pii/S0169433207007386

[29] Brånemark R, Emanuelsson L, Palmquist A, Thomsen P. Bone response to laser-induced micro- and nano-size titanium surface features. Nanomed Nanotechnol Biol Med [Internet]. 2011 [cited 2016 Feb 8];7(2):220–7. Available from: http://www.ncbi.nlm.nih.gov/pubmed/21059406

[30] Györgyey Á, Ungvári K, Kecskeméti G, Kopniczky J, Hopp B, Oszkó A, et al. Attachment and proliferation of human osteoblast-like cells (MG-63) on laser-ablated titanium implant material. Mater Sci Eng C [Internet]. 2013 [cited 2016 Feb 8];33(7):4251–9. Available from: http://www.ncbi.nlm.nih.gov/pubmed/23910340

[31] Surmeneva MA, Surmenev RA, Tyurin AI, Mukhametkaliyev TM, Teresov AD, Koval NN, et al. Comparative study of the radio-frequency magnetron sputter deposited CaP films fabricated onto acid-etched or pulsed electron beam-treated titanium. Thin Solid Films [Internet]. 2014 [cited 2016 Feb 8];571:218–24. Available from: http://www.sciencedirect.com/science/article/pii/S0040609014010104

[32] Zhang XD, Hao SZ, Li XN, Dong C, Grosdidier T. Surface modification of pure titanium by pulsed electron beam. Appl Surf Sci [Internet]. 2011 [cited 2016 Feb 8];257(13):5899–902. Available from: http://www.sciencedirect.com/science/article/pii/S0169433211001723

[33] Cha S, Park Y-S. Plasma in dentistry. Clin Plasma Med. 2015;1:4–7.

[34] Kim J-H, Lee M-A, Han G-J, Cho B-H. Plasma in dentistry: a review of basic concepts and applications in dentistry. Acta Odontol Scand [Internet]. 2014 [cited 2016 Feb 8]; 72(1):1–12. Available from: http://www.ncbi.nlm.nih.gov/pubmed/24354926

[35] Coelho PG, Giro G, Teixeira HS, Marin C, Witek L, Thompson VP, et al. Argon-based atmospheric pressure plasma enhances early bone response to rough titanium surfaces. J Biomed Mater Res A [Internet]. 2012 [cited 2016 Feb 8];100(7):1901–6. Available from: http://www.ncbi.nlm.nih.gov/pubmed/22492543

[36] Lee J-H, Kim Y-H, Choi E-H, Kim K-M, Kim K-N. Air atmospheric-pressure plasma-jet treatment enhances the attachment of human gingival fibroblasts for early peri-implant soft tissue seals on titanium dental implant abutments. Acta Odontol Scand

[Internet]. 2015 [cited 2016 Feb 8];73(1):67–75. Available from: http://www.ncbi.nlm.nih.gov/pubmed/25183251

[37] Yoo E-M, Uhm S-H, Kwon J-S, Choi H-S, Choi EH, Kim K-M, et al. The study on inhibition of planktonic bacterial growth by non-thermal atmospheric pressure plasma jet treated surfaces for dental application. J Biomed Nanotechnol [Internet]. 2015 [cited 2016 Feb 8];11(2):334–41. Available from: http://www.ncbi.nlm.nih.gov/pubmed/26349309

[38] Yang C-H, Li Y-C, Tsai W-F, Ai C-F, Huang H-H. Oxygen plasma immersion ion implantation treatment enhances the human bone marrow mesenchymal stem cells responses to titanium surface for dental implant application. Clin Oral Implants Res [Internet]. 2015 [cited 2016 Feb 8];26(2):166–75. Available from: http://www.ncbi.nlm.nih.gov/pubmed/24313899

[39] Wang X-J, Liu H-Y, Ren X, Sun H-Y, Zhu L-Y, Ying X-X, et al. Effects of fluoride-ion-implanted titanium surface on the cytocompatibility in vitro and osseointegatation in vivo for dental implant applications. Colloids Surf B Biointerfaces [Internet]. 2015 1 [cited 2016 Feb 8];136:752–60. Available from: http://www.ncbi.nlm.nih.gov/pubmed/26519937

[40] Cheng M, Qiao Y, Wang Q, Jin G, Qin H, Zhao Y, et al. Calcium plasma implanted titanium surface with hierarchical microstructure for improving the bone formation. ACS Appl Mater Interfaces [Internet]. 2015 [cited 2016 Feb 8];7(23):13053–61. Available from: http://www.ncbi.nlm.nih.gov/pubmed/26020570

[41] Liang Y, Xu J, Chen J, Qi M, Xie X, Hu M. Zinc ion implantation–deposition technique improves the osteoblast biocompatibility of titanium surfaces. Mol Med Rep [Internet]. 2015 [cited 2016 Feb 8];11(6):4225–31. Available from: http://www.pubmedcentral.nih.gov/articlerender.fcgi?artid=4394954&tool=pmcentrez&rendertype=abstract

[42] Surmenev RA. A review of plasma-assisted methods for calcium phosphate-based coatings fabrication. Surf Coat Technol [Internet]. 2012 [cited 2015 Nov 18];206(8–9):2035–56. Available from: http://www.sciencedirect.com/science/article/pii/S025789721101125X

[43] Xie C, Lu H, Li W, Chen F-M, Zhao Y-M. The use of calcium phosphate-based biomaterials in implant dentistry. J Mater Sci Mater Med [Internet]. 2012 [cited 2016 Feb 8];23(3):853–62. Available from: http://www.ncbi.nlm.nih.gov/pubmed/22201031

[44] Mandracci P, Mussano F, Rivolo P, Carossa S. Surface treatments and functional coatings for biocompatibility improvement and bacterial adhesion reduction in dental implantology. Coatings [Internet]. Multidisciplinary Digital Publishing Institute; 2016 [cited 2016 Feb 8];6(1):7. Available from: http://www.mdpi.com/2079-6412/6/1/7/htm

[45] Damen JJ, Ten Cate JM, Ellingsen JE. Induction of calcium phosphate precipitation by titanium dioxide. J Dent Res [Internet]. 1991 [cited 2016 Feb 8];70(10):1346–9. Available from: http://www.ncbi.nlm.nih.gov/pubmed/1939827

[46] Ellingsen J. A study on the mechanism of protein adsorption to TiO_2. Biomaterials [Internet]. 1991 [cited 2016 Feb 8];12(6):593–6. Available from: http://www.sciencedirect.com/science/article/pii/014296129190057H

[47] Golec TS, Krauser JT. Long-term retrospective studies on hydroxyapatite coated endosteal and subperiosteal implants. Dent Clin N Am [Internet]. 1992 [cited 2016 Feb 8];36(1):39–65. Available from: http://www.ncbi.nlm.nih.gov/pubmed/1310661

[48] Block MS, Kent JN. Prospective review of integral implants. Dent Clin N Am [Internet]. 1992 [cited 2016 Feb 8];36(1):27–37. Available from: http://www.ncbi.nlm.nih.gov/pubmed/1310660

[49] Yukna RA. Placement of hydroxyapatite-coated implants into fresh or recent extraction sites. Dent Clin N Am [Internet]. 1992 [cited 2016 Feb 8];36(1):97–115. Available from: http://www.ncbi.nlm.nih.gov/pubmed/1310664

[50] Tinsley D, Watson CJ, Russell JL. A comparison of hydroxylapatite coated implant retained fixed and removable mandibular prostheses over 4 to 6 years. Clin Oral Implants Res [Internet]. 2001 [cited 2016 Jan 11];12(2):159–66. Available from: http://www.ncbi.nlm.nih.gov/pubmed/11251666

[51] Lee JJ, Rouhfar L, Beirne OR. Survival of hydroxyapatite-coated implants: a meta-analytic review. J Oral Maxillofac Surg [Internet]. 2000 [cited 2016 Feb 8];58(12):1372–9; discussion 1379–80. Available from: http://www.ncbi.nlm.nih.gov/pubmed/11117685

[52] Chang YL, Lew D, Park JB, Keller JC. Biomechanical and morphometric analysis of hydroxyapatite-coated implants with varying crystallinity. J Oral Maxillofac Surg [Internet]. 1999 [cited 2016 Feb 8];57(9):1096–108; discussion 1108–9. Available from: http://www.ncbi.nlm.nih.gov/pubmed/10484111

[53] Wheeler SL. Eight-year clinical retrospective study of titanium plasma-sprayed and hydroxyapatite-coated cylinder implants. Int J Oral Maxillofac Implants 1996 May–Jun; 11(3):340–50.

[54] Giavaresi G, Fini M, Cigada A, Chiesa R, Rondelli G, Rimondini L, et al. Mechanical and histomorphometric evaluations of titanium implants with different surface treatments inserted in sheep cortical bone. Biomaterials [Internet]. 2003 [cited 2016 Feb 8];24(9):1583–94. Available from: http://www.ncbi.nlm.nih.gov/pubmed/12559818

[55] Tsui YC, Doyle C, Clyne TW. Plasma sprayed hydroxyapatite coatings on titanium substrates. Part 1: mechanical properties and residual stress levels. Biomaterials [Internet]. 1998 [cited 2016 Jan 21];19(22):2015–29. Available from: http://www.ncbi.nlm.nih.gov/pubmed/9870753

[56] Tsui YC, Doyle C, Clyne TW. Plasma sprayed hydroxyapatite coatings on titanium substrates. Part 2: optimisation of coating properties. Biomaterials [Internet]. 1998 [cited

2016 Feb 8];19(22):2031–43. Available from: http://www.ncbi.nlm.nih.gov/pubmed/9870754

[57] Aoki H. Science and Medical Applications of Hydroxyapatite. Takayama Press System Center Co. Inc.; Tokyo; 1991.

[58] Albrektsson T, Wennerberg A. Oral implant surfaces: part 1—review focusing on topographic and chemical properties of different surfaces and in vivo responses to them. Int J Prosthodont [Internet]. 2004 [cited 2016 Feb 1];17(5):536–43. Available from: http://www.ncbi.nlm.nih.gov/pubmed/15543910

[59] Bobyn JD, Pilliar RM, Cameron HU, Weatherly GC. The optimum pore size for the fixation of porous-surfaced metal implants by the ingrowth of bone. Clin Orthop Relat Res [Internet]. [cited 2016 Jan 7];(150):263–70. Available from: http://www.ncbi.nlm.nih.gov/pubmed/7428231

[60] Coelho PG, Granjeiro JM, Romanos GE, Suzuki M, Silva NRF, Cardaropoli G, et al. Basic research methods and current trends of dental implant surfaces. J Biomed Mater Res Part B Appl Biomater. 2009;88(2):579–96.

[61] Wennerberg A, Albrektsson T. On implant surfaces: a review of current knowledge and opinions. Int J Oral Maxillofac Implants 2010 Jan–Feb;25(1):63–74.

[62] Mendonça G, Mendonça DBS, Aragão FJL, Cooper LF. Advancing dental implant surface technology—from micron- to nanotopography. Biomaterials [Internet]. 2008 [cited 2015 Sep 24];29(28):3822–35. Available from: http://www.ncbi.nlm.nih.gov/pubmed/18617258

[63] Mendonça G, Mendonça DBS, Simões LGP, Araújo AL, Leite ER, Duarte WR, et al. The effects of implant surface nanoscale features on osteoblast-specific gene expression. Biomaterials [Internet]. 2009 [cited 2016 Feb 8];30(25):4053–62. Available from: http://www.ncbi.nlm.nih.gov/pubmed/19464052

[64] Vetrone F, Variola F, Tambasco de Oliveira P, Zalzal SF, Yi J-H, Sam J, et al. Nanoscale oxidative patterning of metallic surfaces to modulate cell activity and fate. Nano Lett [Internet]. 2009 [cited 2016 Feb 8];9(2):659–65. Available from: http://www.ncbi.nlm.nih.gov/pubmed/19159323

[65] Dalby MJ, McCloy D, Robertson M, Agheli H, Sutherland D, Affrossman S, et al. Osteoprogenitor response to semi-ordered and random nanotopographies. Biomaterials [Internet]. 2006 [cited 2016 Jan 11];27(15):2980–7. Available from: http://www.ncbi.nlm.nih.gov/pubmed/16443268

[66] Dalby MJ, McCloy D, Robertson M, Wilkinson CDW, Oreffo ROC. Osteoprogenitor response to defined topographies with nanoscale depths. Biomaterials [Internet]. 2006 [cited 2016 Feb 8];27(8):1306–15. Available from: http://www.ncbi.nlm.nih.gov/pubmed/16143393

[67] Bucci-Sabattini V, Cassinelli C, Coelho PG, Minnici A, Trani A, Dohan Ehrenfest DM. Effect of titanium implant surface nanoroughness and calcium phosphate low impreg-

nation on bone cell activity in vitro. Oral Surg Oral Med Oral Pathol Oral Radiol Endod [Internet]. 2010 [cited 2016 Feb 8];109(2):217–24. Available from: http://www.ncbi.nlm.nih.gov/pubmed/20031453

[68] Bjursten LM, Rasmusson L, Oh S, Smith GC, Brammer KS, Jin S. Titanium dioxide nanotubes enhance bone bonding in vivo. J Biomed Mater Res A [Internet]. 2010 [cited 2016 Feb 8];92(3):1218–24. Available from: http://www.ncbi.nlm.nih.gov/pubmed/19343780

[69] Kodama A, Bauer S, Komatsu A, Asoh H, Ono S, Schmuki P. Bioactivation of titanium surfaces using coatings of TiO(2) nanotubes rapidly pre-loaded with synthetic hydroxyapatite. Acta Biomater [Internet]. 2009 [cited 2016 Feb 8];5(6):2322–30. Available from: http://www.ncbi.nlm.nih.gov/pubmed/19332383

[70] de Oliveira PT, Zalzal SF, Beloti MM, Rosa AL, Nanci A. Enhancement of in vitro osteogenesis on titanium by chemically produced nanotopography. J Biomed Mater Res A [Internet]. 2007 [cited 2016 Feb 8];80(3):554–64. Available from: http://www.ncbi.nlm.nih.gov/pubmed/17031821

[71] Rupp F, Gittens RA, Scheideler L, Marmur A, Boyan BD, Schwartz Z, et al. A review on the wettability of dental implant surfaces I: theoretical and experimental aspects. Acta Biomater [Internet]. 2014 [cited 2016 Feb 8];10(7):2894–906. Available from: http://www.pubmedcentral.nih.gov/articlerender.fcgi?artid=4041806&tool=pmcentrez&rendertype=abstract

[72] Buser D, Broggini N, Wieland M, Schenk RK, Denzer AJ, Cochran DL, et al. Enhanced bone apposition to a chemically modified SLA titanium surface. J Dent Res [Internet]. 2004 [cited 2016 Feb 3];83(7):529–33. Available from: http://www.ncbi.nlm.nih.gov/pubmed/15218041

[73] Fraker AC, Ruff AW, Sung P, Van Orden AC, Speck KM. Surface Preparation and Corrosion Behavior of Titanium Alloys for Surgical Implants. ASTM Special Technical Publication [Internet]. 1983. pp. 206–19. Available from: http://www.scopus.com/inward/record.url?eid=2-s2.0-0020905617&partnerID=tZOtx3y1

[74] Wei Xia, Carl Lindahl, Jukka Lausmaa and Ha°kan Engqvist (2011). Biomimetic Hydroxyapatite Deposition on Titanium Oxide Surfaces for Biomedical Application, Advances in Biomimetics, Prof. Marko Cavrak (Ed.), InTech, DOI: 10.5772/14900. Available from: http://www.intechopen.com/books/advances-in-biomimetics/biomimetic-hydroxyapatite-deposition-on-titanium-oxide-surfaces-for-biomedical-application

[75] Jarmar T, Palmquist A, Brånemark R, Hermansson L, Engqvist H, Thomsen P. Characterization of the surface properties of commercially available dental implants using scanning electron microscopy, focused ion beam, and high-resolution transmission electron microscopy. Clin Implant Dent Relat Res [Internet]. 2008 [cited 2016 Feb 8];10(1):11–22. Available from: http://www.ncbi.nlm.nih.gov/pubmed/18254738

[76] Sul Y-T, Byon E, Wennerberg A. Surface characteristics of electrochemically oxidized implants and acid-etched implants: surface chemistry, morphology, pore configurations, oxide thickness, crystal structure, and roughness. Int J Oral Maxillofac Implants 2008 Jul–Aug;23(4):631–40.

[77] Sawase T, Jimbo R, Wennerberg A, Suketa N, Tanaka Y, Atsuta M. A novel characteristic of porous titanium oxide implants. Clin Oral Implants Res [Internet]. 2007 [cited 2016 Feb 8];18(6):680–5. Available from: http://www.ncbi.nlm.nih.gov/pubmed/17868377

[78] Coelho PG, Lemons JE. Physico/chemical characterization and in vivo evaluation of nanothickness bioceramic depositions on alumina-blasted/acid-etched Ti–6Al–4V implant surfaces. J Biomed Mater Res A [Internet]. 2009 [cited 2016 Feb 8];90(2):351–61. Available from: http://www.ncbi.nlm.nih.gov/pubmed/18508352

[79] He J, Zhou W, Zhou X, Zhong X, Zhang X, Wan P, et al. The anatase phase of nanotopography titania plays an important role on osteoblast cell morphology and proliferation. J Mater Sci Mater Med [Internet]. 2008 [cited 2016 Feb 8];19(11):3465–72. Available from: http://www.ncbi.nlm.nih.gov/pubmed/18592349

[80] Sollazzo V, Pezzetti F, Scarano A, Piattelli A, Massari L, Brunelli G, et al. Anatase coating improves implant osseointegration in vivo. J Craniofac Surg [Internet]. 2007 [cited 2016 Feb 8];18(4):806–10. Available from: http://www.ncbi.nlm.nih.gov/pubmed/17667669

[81] Chen CA, Huang YS, Chung WH, Tsai DS, Tiong KK. Raman spectroscopy study of the phase transformation on nanocrystalline titania films prepared via metal organic vapour deposition. J Mater Sci Mater Electron [Internet]. 2008 [cited 2016 Feb 8];20(S1): 303–6. Available from: http://link.springer.com/10.1007/s10854-008-9595-3

[82] Ocana M, Garcia-Ramos JV, Serna CJ. Low-Temperature nucleation of rutile observed by Raman spectroscopy during crystallization of TiO_2. J Am Ceram Soc [Internet]. 1992 [cited 2016 Feb 8];75(7):2010–2. Available from: http://doi.wiley.com/10.1111/j.1151-2916.1992.tb07237.x

[83] Yin H, Wada Y, Kitamura T, Kambe S, Murasawa S, Mori H, et al. Hydrothermal synthesis of nanosized anatase and rutile TiO_2 using amorphous phase TiO_2. J Mater Chem [Internet]. The Royal Society of Chemistry; 2001 [cited 2016 Feb 8];11(6):1694–703. Available from: http://pubs.rsc.org/en/content/articlehtml/2001/jm/b008974p

[84] Lim YJ, Oshida Y, Andres CJ, Barco MT. Surface characterizations of variously treated titanium materials. Int J Oral Maxillofac Implants 2001 May–Jun;16(3):333–42.

[85] Rupp F, Haupt M, Eichler M, Doering C, Klostermann H, Scheideler L, et al. Formation and photocatalytic decomposition of a pellicle on anatase surfaces. J Dent Res [Internet]. 2012 [cited 2016 Feb 8];91(1):104–9. Available from: http://www.ncbi.nlm.nih.gov/pubmed/21979134

[86] Jarmar T, Palmquist A, Brånemark R, Hermansson L, Engqvist H, Thomsen P. Technique for preparation and characterization in cross-section of oral titanium implant surfaces using focused ion beam and transmission electron microscopy. J Biomed Mater Res A [Internet]. 2008 [cited 2016 Feb 8];87(4):1003–9. Available from: http://www.ncbi.nlm.nih.gov/pubmed/18257067

[87] Huang N, Chen YR, Luo JM, Yi J, Lu R, Xiao J, et al. In vitro investigation of blood compatibility of Ti with oxide layers of rutile structure. J Biomater Appl [Internet]. 1994 [cited 2016 Feb 8];8(4):404–12. Available from: http://www.ncbi.nlm.nih.gov/pubmed/8064591

[88] Wei Xia, Carl Lindahl, Jukka Lausmaa and Ha°kan Engqvist (2011). Biomimetic Hydroxyapatite Deposition on Titanium Oxide Surfaces for Biomedical Application, Advances in Biomimetics, Prof. Marko Cavrak (Ed.), ISBN: 978–953–307–191–6, InTech, Available from: http://www.intechopen.com/books/advances-inbiomimetics/biomimetic-hydroxyapatite-deposition-on-titanium-oxide-surfaces-for-biomedical-application

[89] Gaintantzopoulou M, Zinelis S, Silikas N, Eliades G. Micro-Raman spectroscopic analysis of TiO_2 phases on the root surfaces of commercial dental implants. Dent Mater [Internet]. 2014 [cited 2016 Feb 8];30(8):861–7. Available from: http://www.ncbi.nlm.nih.gov/pubmed/24946981

Chapter 3

Functional Metal Oxide Thin Films Grown by Pulsed Laser Deposition

Mihaela Filipescu, Alexandra Palla Papavlu and Maria Dinescu

Additional information is available at the end of the chapter

http://dx.doi.org/10.5772/62986

Abstract

The aim of this work is to show that material processing by laser-based technologies can lead to the growth of multifunctional thin films with potential in a large area of applications. The synthesis of Hf, Ta, Si, and Al metal oxides described here relies on the use of pulsed laser deposition (PLD), or radiofrequency (RF) assisted PLD. The morphology and structure of the as-grown thin films are investigated by atomic force microscopy, X-ray diffraction, and transmission electron microscopy, whilst the optical properties are determined by spectroellipsometry. The dielectric behaviour of the deposited layers is investigated by electrical measurements.

It is shown that by tuning the deposition parameters, the materials of interest can be synthesized as compact and dense oxide layers. Parameters such as substrate temperature, oxygen pressure, or laser wavelength have a critical impact on the crystallinity of the films, as well as on the characteristic functional properties. For example, hafnium dioxide, tantalum oxide, and aluminium oxide layers grown by RF-PLD at room temperature have low leakage currents, which make them useful for dielectric gates. When high substrate temperatures are involved in the PLD process, these oxide layers have a crystalline structure and smooth surfaces, with potential in antireflective coatings.

Keywords: metal oxides, thin films, pulsed laser deposition, radiofrequency assisted PLD, antireflective coatings, high-k dielectrics

1. Introduction

In the past century, the continuous development of human society led to exceptional scientific and technological accomplishments, miniaturization being a key work. Challenges such as the decrease in the production costs together with the transition from the micro to the nanoscale represent, at present, the focus of the international scientific and industrial community. Metal oxide materials (in particular Hf, Al, Si, and Ta oxides) have been attracting considerable attention, due to their complexity which originates from their seemingly antagonistic properties that make them attractive in different applications [1]. Hafnium dioxide and aluminium oxide processed as smooth films have been considered for applications such as dielectric mirrors and high-k candidates for next-generation MOS gates [2]; grown as nanostructured materials, they show the potential for biological applications. Tantalum oxide as a smooth thin film had a fast evolution as an important material in opto-electronics, being extensively used in the production of capacitors and anti-reflection coatings [3]. The nanostructured surface morphology of tantalum oxide makes it suitable for solid-state oxygen sensors and thin-film catalysts [4, 5].

The metallic oxides can be obtained as thin films by numerous techniques, such as plasma enhanced chemical vapour deposition [6], chemical vapour deposition [7], atomic layer deposition [8], radiofrequency (RF) sputtering [9], etc. Major disadvantages of these techniques are related to the high processes temperatures, the use of expensive and highly corrosive precursors, and subsequent thermal treatments for crystallization. A reliable method for obtaining thin films of simple or complex compounds is pulsed laser deposition (PLD), also called laser ablation.

PLD has gained worldwide acknowledgement as a reliable method for obtaining thin films of simple or complex materials. The PLD relies on the interaction of the laser beam with a target material (solid or liquid), thus producing a plasma plume through which the material is carried on a substrate, as a thin film. This method has several advantages: (a) the deposition chamber is a "clean" reactor as the energy source (laser) is outside the reaction chamber, (b) the deposition process can be easily controlled because the processes involved are strongly influenced by the laser parameters (wavelength, laser fluence, laser spot area, laser pulse duration, repetition rate, etc.), (c) the thickness of the film can be controlled by the number of pulses that irradiate the material, (d) the laser radiation can be focused/imaged in a very small spot which allows the selective processing regions of interest, (e) the transfer of the target to the substrate can produce new materials in metastable states, that are impossible to be achieved by other techniques.

A general problem when depositing metal oxide thin films is the appearance of oxygen vacancies inside the layer and at the layer-substrate interface. A possible solution to this problem is the usage of a hybrid technique that combines the advantages of conventional PLD with the addition of a RF oxygen gas stream directed toward the substrate. During the growth process, the species from the laser plasma plume (Hf, Ta, Al, or Si oxides, ions, etc.) interact with the excited and/or ionized oxygen species generated by the RF discharge. This complex developed deposition technique is named as RF assisted pulsed laser deposition (RF-PLD) [10].

The plasma beam source consists of a double chamber discharge system supplied by a RF power supply. In the active chamber, the discharge is generated in an oxygen flow between two parallel electrodes. It expands into the ablation chamber as a plasma beam, through an aperture drilled in the bottom electrode, which acts as a nozzle. RF-PLD is suitable for the synthesis of smooth and compact thin films of metal oxides or nanostructured thin films.

Therefore, in the following sections, the deposition of various metal oxide materials (Si, Ta, Hf, and Al) as thin films by PLD and hybrid PLD (RF assisted), with an emphasis on high-k gate dielectrics and antireflective coatings shall be discussed.

2. Experimental methods: PLD and RF assisted PLD

High-purity metal oxide targets (i.e., Hf, Ta, Si, and Al) were used for the ablation experiments using either an ArF excimer laser working at 193 nm wavelength with a variable repetition rate (1–50 Hz), pulse duration of 30 ns, or a Nd-YAG laser Surellite II, working at 355 or 266 nm, 5–7 ns pulse duration, and 1–10 Hz repetition rate. The laser fluence was optimized for each material and the target-substrate distance was 4 cm.

All substrates used for thin-film growth were thoroughly cleaned in successive ultrasonic bath (acetone, ethanol, and deionized water) to remove surface contamination before being inserted into the deposition chamber. Prior to the depositions, the vacuum chamber was evacuated with a turbomolecular pump to pressures below 10^{-5} mbar. Oxygen gas flow during the depositions was adjusted by mass flow controllers. The substrates were heated during the depositions at temperatures between 100–600°C.

To the classic PLD setup, a plasma beam was added in order to enhance the reactivity on the substrate due to the presence of an excited and ionized beam of oxygen atoms and molecules produced by the RF discharge. The plasma beam source consisted of a double chamber discharge system supplied by a RF (13.56 MHz, CESAR 1310, RF maximum power 1000 W) power supply. In the active chamber, the discharge was generated in a flow of oxygen between two parallel electrodes. The oxygen gas stream expanded into the ablation chamber through a nozzle of ~2 mm in diameter drilled in the bottom electrode. The discharge power was 100 W and the distance between the RF discharge gun and the substrate was 4 cm.

Details on the experimental deposition parameters used for processing the metal oxides in this work are given in the corresponding sections.

3. Thin-film characterization

The morphology of the as-grown metal oxide thin films was analyzed by atomic force microscopy (AFM) with a XE-100 AFM from Park Systems, on different areas and dimensions. All the deposited thin films were scanned in non-contact mode. Transmission electron microscopy (TEM) images were recorded on a JEOL ARM 200F analytical high-resolution electron microscope operating at 200 kV primary accelerating voltage after gently scraping the deposited films and collecting the product on carbon-covered copper grids.

Spectroellipsometry data were obtained in the 300–1000 nm spectral range at an incidence angle of 70° using a Woollam V-VASE apparatus equipped with a HS-190 monochromator.

Changes in the polarization state of an incident beam were analyzed, from linearly polarized to elliptical. These changes can be expressed by two parameters: Ψ (amplitude ratio) and Δ (phase difference) [11]. The experimental values for Ψ and Δ are acquired in the spectral range of 400–1000 nm, with a step of 2 nm, and at the beam incident at an angle of 70°. Since ellipsometry is a comparative technique, an optical model was required in order to generate the theoretical curves of Ψ and Δ, which are to be fitted to the experimental data. For our films, we assumed an optical model consisting of four material layers: the silicon substrate, the native silicon oxide, the thin film of the metallic oxide layer, and a top rough layer. The values of the optical constants for silicon and native silicon oxide were taken from literature [12]. The top rough layer was considered to be a mix of materials, 50% oxide layer and 50% air (voids), in Bruggeman approximation [11].

X-ray measurements conducted in a Bragg-Brentano geometry (θ–2θ, between 20° and 100°) were carried out on a PANalytical X'Pert MRD system (CuKα, λ = 1.5418 Å).

To evaluate the leakage currents through the thin films, metallic contacts were deposited as top electrodes with 100 nm thickness. The current-voltage characteristic was measured using a Keithley instrument with a precision on the order of nano-amperes.

4. Silicon dioxide thin films

Nowadays, the electronic industry is based on silicon dioxide technology. Since the 1980s, SiO_2 is used as gate dielectric material for complementary metal-oxide-semiconductor (CMOS) technology. The technological requirements include high speed, low static power, and a wide range of power supply and output voltages [13]. The miniaturization trend, i.e. from the micro to the nano-scale resulted in the fabrication of the field effect transistor (FET), which lead to a high expansion of the technology and communication markets. Therefore, the metal-oxide-semiconductor field effect transistor (MOSFET) based on silicon dioxide gain enormous importance in the development of electronic devices.

Amorphous SiO_2 thin films are thermodynamically and electrically stable, having a good quality of the interface with the Si wafer, and superior electrical isolation properties. In the food industry, silicon dioxide is a common additive, used as a flow agent in powdered foods, or for water adsorption in hygroscopic applications. Furthermore, SiO_2 is a good refractory material, used as antireflective coating for laser optical components.

In this work we focus on the deposition of SiO_2 thin films to be used in antireflective coating applications. For this application, it is important to obtain featureless SiO_2 thin films, i.e. free of pores and droplets.

SiO_2 thin films are deposited either by irradiation with a Nd:YAG laser (operated at 266 nm) or an ArF laser (193 nm) of a silicon target (classical PLD). The thin-film depositions are carried out in an oxygen atmosphere (0.01–0.1 mbar) using a laser fluence of 1.5 J/cm^2. The silicon substrates used as collectors are heated at 600°C during the depositions, and are set parallel to the target, at 4 cm distance.

For the applications envisioned here, i.e. antireflective coatings, the thin-film surface morphology characterization is mandatory. It has been found [14, 15] that the laser wavelength plays an important role in the ablation process, influencing the thin film properties. The films grown from the target irradiated with 193 nm laser wavelength, at 0.01 mbar oxygen pressure, present micrometric droplets, with high roughness (RMS) of ~40 nm (see **Figure 1(a)**). In contrast, the SiO_2 thin films deposited from the target irradiated with 266 nm laser wavelength reveal both a reduced number of droplets and roughness (~30 nm). Increased oxygen pressure during thin-film deposition (0.1 mbar) together with irradiation at 266 nm laser wavelength lead to a considerable decrease of the films' roughness (below 10 nm) (see **Figure 1(b)**).

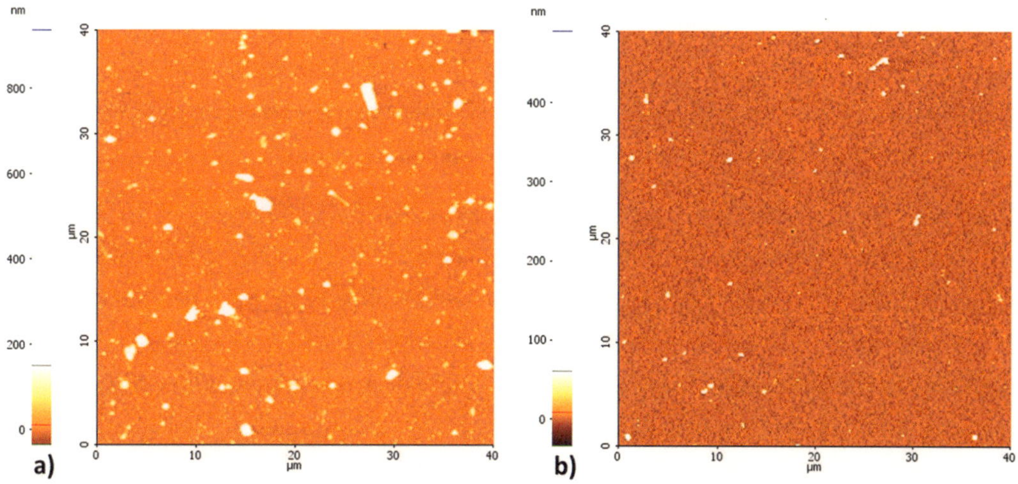

Figure 1. AFM images for thin films of silicon oxide grown by PLD at (a) 0.01 mbar of oxygen and λ = 193 nm, (b) 0.1 mbar of oxygen and λ = 266 nm.

Furthermore, the electron diffraction pattern together with the HRTEM image of the SiO_2 thin films deposited at 266 nm is shown in **Figure 2**. The electron diffraction patterns shown in **Figure 2(a)** reveal strong diffraction maxima attributed to the Si(100) substrate and weak diffraction maxima corresponding to the SiO_2 cubic structure, spatial group SG:$P2_13$ and lattice constant a = 0.716 nm (JCPDS file 85-0621). The HRTEM image in **Figure 2(b)** reveals a columnar growth of the SiO_2 thin films. The differences of hill/valley height are small, below 2 nm. It is noticeable that the columns have almost the same width from the base to the top of the film. The columns differ between them only by the crystallites orientation. At the Si substrate-SiO_2 thin film interface, there is an amorphous SiO_2 thin layer of 2 nm thickness.

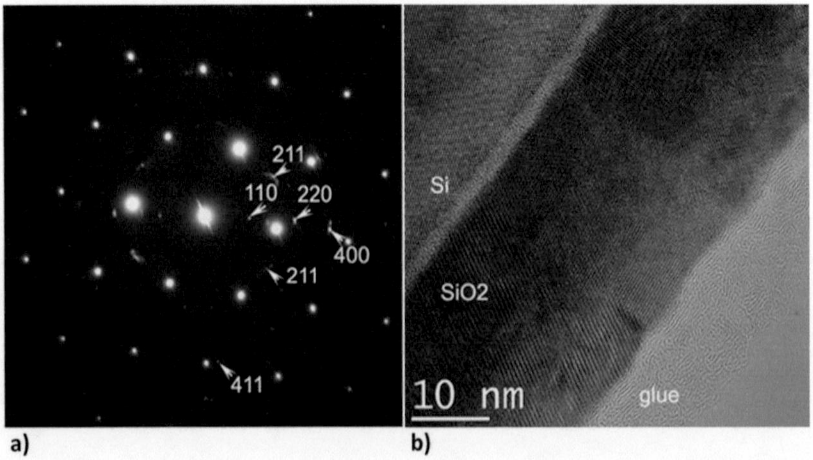

Figure 2. (a) Electron diffraction pattern of the SiO_2 thin film deposited onto a Si substrate, (b) cross-section HRTEM of the SiO_2 thin film deposited on a Si substrate. The laser processing wavelength was 266 nm.

The SiO_2 thin films deposited by PLD at different laser wavelengths (193 and 266 nm) are investigated by spectroellipsometry to determine the refractive index.

It has been found that the laser wavelength used to process the thin films plays an important role in the optical properties of the thin films. Using the 266 nm wavelength, thin films with a refractive index $n \sim 1.45$, i.e. similar to the values from the data base (reference) are obtained. For 193 nm laser wavelength irradiation, the deposited thin films have a higher $n \sim 1.55$, which can be explained by the presence of a higher amount of Si in the SiO_2 thin films (**Figure 3**).

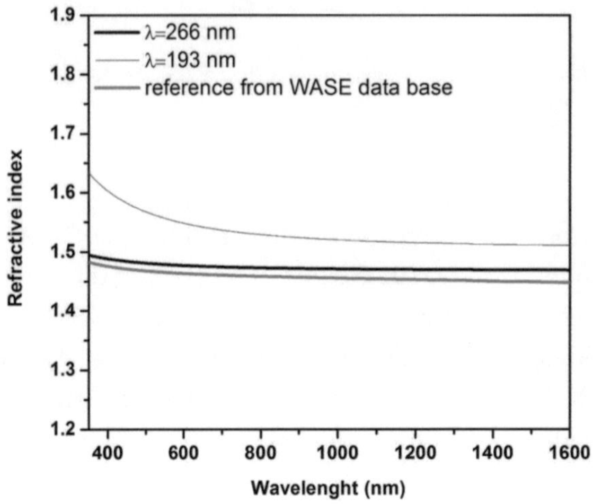

Figure 3. Refractive index of the SiO_2 thin films deposited by PLD compared to a reference SiO_2 thin film.

http://dx.doi.org/10.5772/62986

5. Tantalum oxide

In the microelectronics field, higher integrated circuit functionality and performance at lower cost are continuous demands. The requirement of an increased circuit density leads to a higher density of transistors on a wafer. SiO_2 has some limitations, for example, high leakage current, therefore new materials are studied. Replacing SiO_2 as gate dielectric in metal-oxide-semiconductor devices with high-k dielectric materials avoids the decrease in the dielectric's thickness to dimensions of angstroms, keeping it in the nanometres range [16].

Tantalum pentoxide (Ta_2O_5) quickly evolved as an important material in microelectronics. Due to its high dielectric constant (around 25) and to its chemical and thermal stability, it is a promising candidate as dielectric material in dynamic random-access memory devices. It can be used in microelectronics processing, replacing the conventional thin films of SiO_2 or Si_3N_4/SiO_2. The high refractive index and low absorption coefficient on wide visible range make it suitable in optical devices as antireflective coating for solar cells based on silicon, and as material for waveguides.

Other applications of tantalum oxide include coatings for mirrors, and biological and chemical sensors. Ta_2O_5 can be used as gate insulator in thin films and MOSFET. Moreover, Ta_2O_5 has a high refractive index (n ~ 2.2 at 633 nm), a wide band gap (*e.g.* ~ 4.2 eV), and free absorption for wavelengths in the 300 nm to 2.0 μm range allowing to be used in "charge coupled" devices. Furthermore, due to its chemical stability, Ta_2O_5 can be used as protective layer to corrosion.

For applications in electronics as gate dielectric, the morphological and electrical properties of Ta_2O_5 thin films exhibit high interest. The films roughness should be in the order of tens of nanometres and leakage currents should be very low.

Thin layers of Ta_2O_5 are prepared by PLD and RF plasma-assisted PLD [17]. The laser beam (266 or 355 nm) is focused on a tantalum metallic target to obtain Ta_2O_5 thin films. The distance between the target and the Si or Pt-coated Si (Pt/Si) substrates is 4 cm. The substrates are heated to 400 or 600°C. Depositions are carried out in oxygen atmosphere, at a constant pressure of 0.05 mbar. The laser fluence is varied between 1–3 J/cm^2 in order to avoid the formation of droplets on the films' surface. Thin films with thicknesses of around 150 nm are obtained as a result of 15,000 laser pulses.

Surface morphology has an important role for the applications presented above, in particular, for gate dielectric and antireflective coatings. The thin films' requirements are low roughness, lack of pores, and uniform thickness over a large area.

First, we investigate the morphology of two thin films deposited by PLD at 1 J/cm^2 and two laser wavelengths, i.e. 266 nm and 355 nm (**Figure 4**). Compact thin films with only a few droplets and low roughness (below 4 nm) are obtained with 266 nm wavelength (**Figure 4a**). In contrast, thin films with micrometric pores and droplets, and roughness around 10 nm are obtained by using 355 nm laser wavelength, as shown in **Figure 4b**. The tantalum oxide thin films are amorphous as revealed by HRTEM investigation (not shown here).

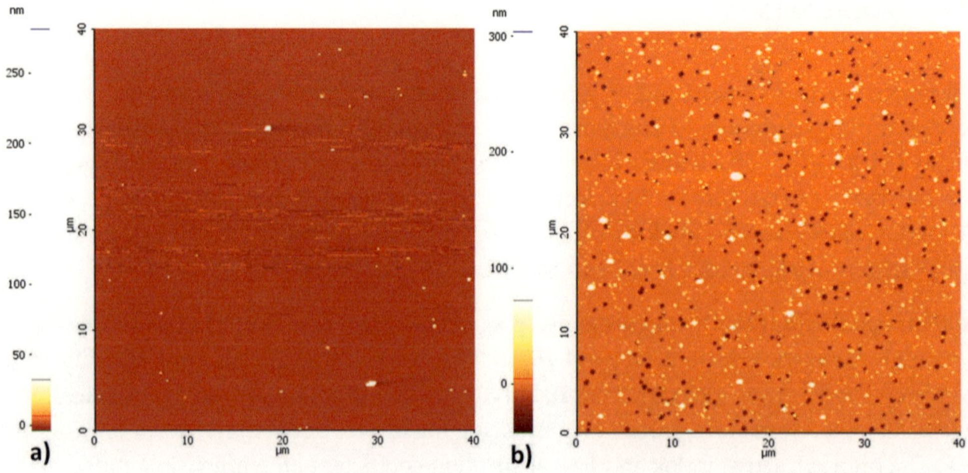

Figure 4. AFM images of Ta_2O_5 thin films grown on Si substrate at different wavelengths: (a) 266 nm, and (b) 355 nm.

Spectroscopic ellipsometry investigations show that the tantalum oxide layers present refractive indices close to those of tantalum pentoxides (Ta_2O_5) from the VASE32 database [18] (**Figure 5**).

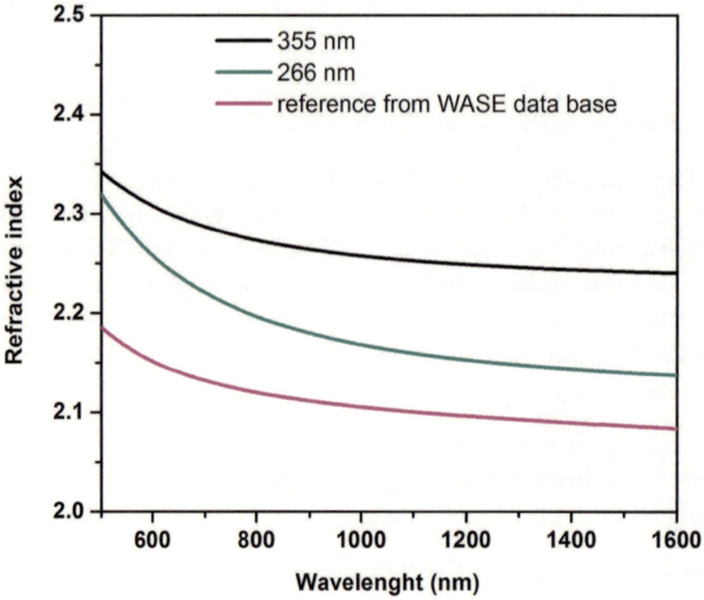

Figure 5. Refractive index of the Ta_2O_5 thin films deposited by PLD compared to a reference Ta_2O_5 thin film.

The I–V characteristics are measured on tantalum oxide layers deposited on Pt/Si substrate with aluminium top electrodes having areas around 0.22–0.23 mm^2. The influence of the wavelength and RF plasma addition to the PLD system on the leakage currents densities for tantalum oxide is studied. The influence of the RF discharge and the wavelength on the electrical properties is highlighted in **Figure 6**.

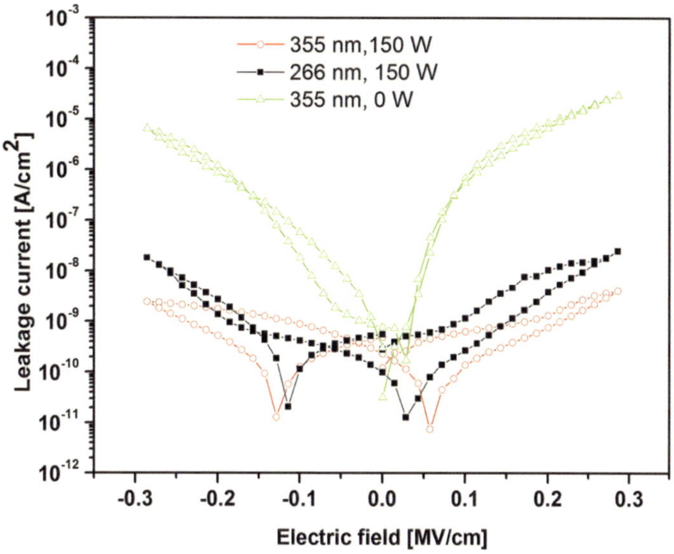

Figure 6. The I–V characteristics of the Ta_2O_5 thin films [17].

The thin films grown by RF-PLD have low leakage currents densities with three orders of magnitude smaller than for samples grown by PLD. The structural ordering induced by RF plasma addition results in a significant decrease in the leakage current densities. Thin films deposited using the third laser harmonic (355 nm) showed one order of magnitude smaller leakage currents, as compared to those obtained using the fourth laser harmonic (266 nm). These low leakage currents densities corroborated with smooth, low roughness thin films made from tantalum oxide grown by RF-PLD, the suitable candidate for dielectric gates. In addition, our thin films have superior dielectric properties (refraction index and leakage current) when compared to results reported in literature by other deposition techniques (see **Table 1**).

Deposition technique	Refraction index (550 nm)	Leakage current density (A/cm²)
RF-PLD (this work)	2.27	4.8×10^{-10}
PLD [19]	2.22	-
Thermal oxidation [20]	2.10	7×10^{-10}
Atomic layer deposition [21]	2.13	1×10^{-8}
Rf magnetron sputtering [22]	2.11	1×10^{-6}

Table 1. Comparison between the refractive index of Ta_2O_5 thin films deposited by different deposition methods.

To sum up, the high refractive index (2.27), low absorption coefficient, and the fact that the material is inert makes it useful as antireflection coating and waveguide material [23, 24].

6. Hafnium oxide

Hafnium dioxide, also called hafnia, is a high-index, low-absorption material used for coatings in the near-UV (below 300 nm) to IR (~10 μm) regions [25]. Typical applications include near-UV laser antireflective coatings and dielectric mirror designs. Adhesion is excellent to glass, quartz, to the most other oxides, and to metals such as aluminium and silver. The refractive index of hafnia thin films changes with respect to the high energy deposition techniques and substrate temperature, because both parameters decrease the void volume by increasing the packing density of the microstructure [26].

The refractive index below 300 nm wavelength is near 2; therefore, hafnia can be combined in multilayers with silicon dioxide (n = 1.48) to form high index/low index-contrast multilayer structures with high laser damage thresholds for UV laser applications [27]. Hard, scratch-free, dense, and adherent coatings can be deposited on relatively low temperature substrates. Its abrasion resistance provides the protection of metal mirrors.

HfO_2 is suitable for gate dielectric due to its stability in direct contact with Si at temperatures required for processing and due to its high dielectric constant of 18–25. For this specific application, thin films of HfO_2 are prepared by RF-PLD on Pt/Si and Si(100) substrates (heated at 100°C) [28]. The ablation of a hafnium target is carried out in reactive atmosphere of oxygen. The laser fluence is set to 5 J/cm^2. And, 12000 laser pulses are shot to obtain HfO_2 thin films with thickness of approximately 150 nm. Different deposition parameters, i.e. laser wavelength (266 or 355 nm) and oxygen pressure (0.5–0.05 mbar) are varied to obtain thin films with smooth morphology and dielectric properties.

It has been found that the laser wavelength does not influence the thin films morphology, only a slight decrease in the surface roughness is being observed in the case of samples deposited at 266 nm. The surface is uniform and compact, with rare droplets (not shown) and roughness below 15 nm [28]. Auger electron spectroscopy reveals thin films with stoichiometry close to HfO_2 (Hf 35%: O 65%).

In addition, AFM investigations show that the variation of oxygen pressure does not influence the surface morphology. The results of electrical measurements on hafnia thin films prepared in the same conditions but with different oxygen pressures indicate small leakage current density values, between 10^6 and 10^3 A/cm^2 on a narrow electric field (~0.2 MV/cm) for low oxygen pressures (0.05 – 0.1 mbar); the thin films grown at a slightly higher oxygen pressure (0.5 mbar) exhibit lower leakage current densities for a wide electric field (up to 0.4 MV/cm) (see **Figure 7**).

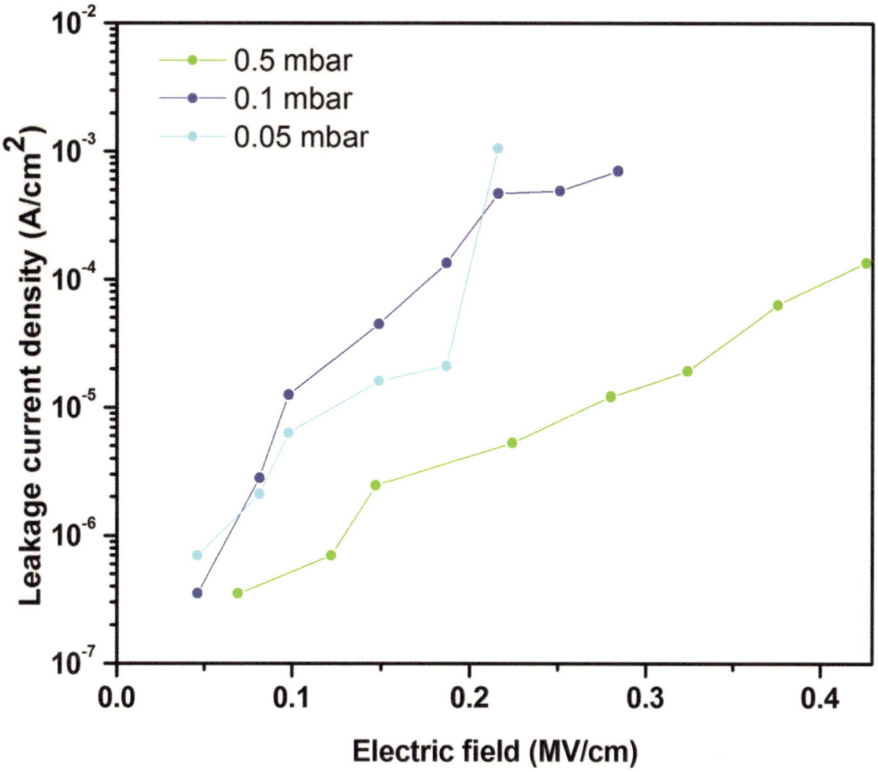

Figure 7. Leakage current density vs. electric field of hafnia thin films prepared by RF-PLD at different oxygen pressures [28].

In conclusion, it has been shown that hafnia thin films grown in low oxygen content (≤ 0.5 mbar) have low leakage currents suitable for applications in MOSFET devices.

Further on, thin films with low surface roughness and high refractive index are candidates for coatings in dielectric mirrors. For this application, thin layers of hafnium oxide are produced in 0.01 mbar of oxygen, by ablation at different wavelengths (193, 266, and 355 nm). The substrates (silicon and Pt/Si) are kept at 600°C during the depositions. The laser fluence is set at 1.5 J/cm^2.

Laser wavelength has an important role in obtaining droplet-free thin films. After target irradiation with 193 nm, the obtained films show a low roughness (below 1 nm) for 20 nm film thickness (**Figure 8a**). Thin films produced in the same experimental conditions using 266 nm laser wavelength have higher roughness (about 6 nm), however, for a higher thickness (227 nm) (not shown). Hafnia target irradiated at higher wavelength (355 nm) leads to thin films with micrometric droplets and conglomerates; the surface roughness is very high (25 nm) for a small film thickness (40 nm) (**Figure 8b**).

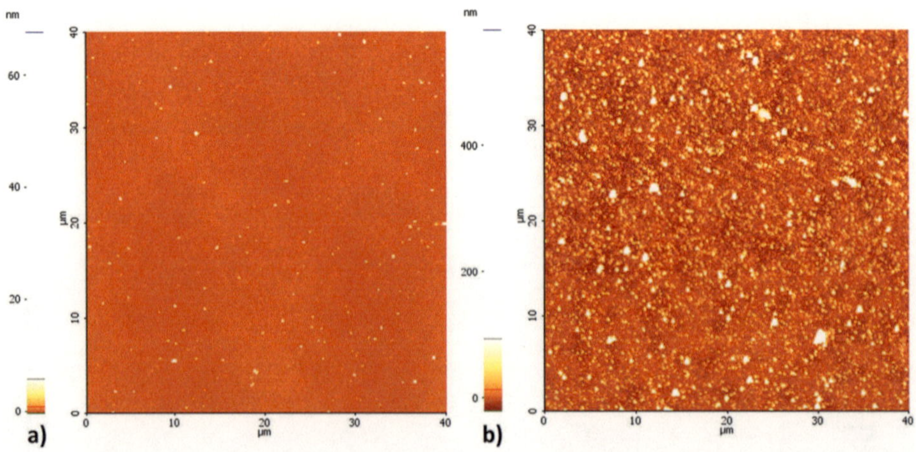

Figure 8. AFM images of hafnia thin films grown on Si substrates at (a) 193 nm and (b) 355 nm laser wavelength.

The structural investigation by XRD shows that the HfO_2 thin films are crystalline with (-111) orientation (PDF card 00-034-0104) (see **Figure 9**).

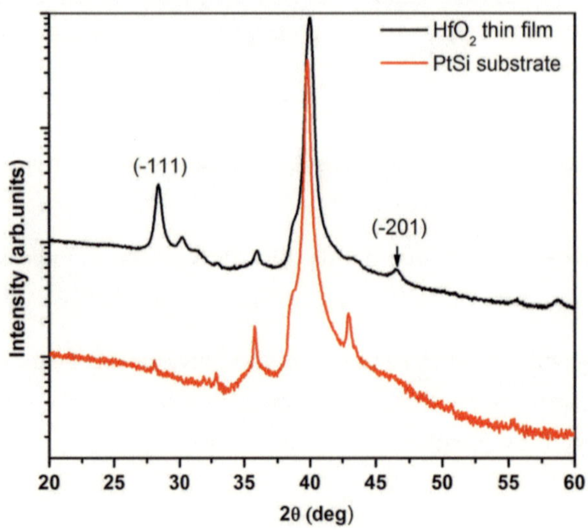

Figure 9. XRD patterns of an as-grown HfO_2 thin film deposited by PLD at 266 nm laser wavelength.

In addition to hafnia surface morphology, the laser wavelength influences the optical behaviour of the hafnia films. Thin films with refractive index values close to the value of HfO_2 reference from WASE data base are obtained at 266 and 355 nm. The irradiation of the target with 193 nm laser wavelength leads to thin films with a higher refractive index, due to the high roughness (voids presence) corroborated with a higher content of Hf in the oxide layer (**Figure 10**). And, 266 nm laser wavelength has been found as the suitable wavelength for smooth (RMS

= 5.8 nm) thin layers with good refractive index (n ~ 2) to be further used in antireflective coating applications.

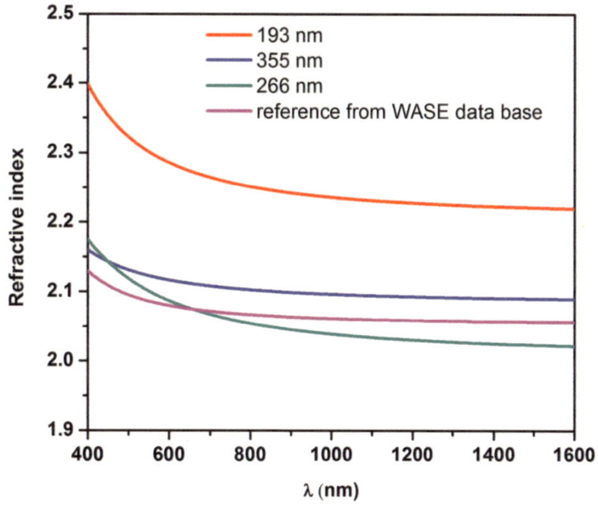

Figure 10. Refractive index of the HfO_2 thin films deposited by PLD compared to a reference HfO_2 thin film.

7. Aluminium oxide

Another dielectric candidate for replacing silicon dioxide in MOS devices is aluminium dioxide (alumina). It has excellent conduction and valence band offsets on contact with Si [29, 30] being suitable for use with n- and p-type Si over a range of doping levels. Due to its high melting point, Al_2O_3 is used as a refractory material; it is a hard dielectric material with a refractive index of 1.77 [31] and presents high corrosion resistance. Aluminium oxide is a good candidate to integrated optics and as antireflection coating [32] due to the possibility to tune its spectral properties [33].

Taking into account the dielectric behaviour, thin films of alumina are deposited by PLD on Ti-coated glass substrates [34]. An alumina target is ablated with 193 nm laser wavelength, in oxygen atmosphere, at a laser fluence of 4 J/cm^2.

Nanostructured thin films, droplet-free, with roughness below 5 nm are revealed by AFM investigation. The permittivity and dielectric losses vs. temperature are measured in the range of –150 to 150°C, in a 42 Hz and 5 MHz frequency range. Thin films with good stability of the permittivity and low loss values are obtained by PLD. For example, at 0°C, the dielectric constant of Al_2O_3 thin films is around 7.2 [34].

The role of aluminium oxide thin films as antireflective coatings is studied from a morphological and optical point of view. In order to obtain thin films suitable for optical applications, the deposition experiments are carried out in oxygen atmosphere (0.01 mbar), at 193 or 266 nm wavelength. The Si substrates are heated at 600°C during depositions.

The laser fluence is set at 6 J/cm^2 for the depositions carried out at 193 nm and 1.5 J/cm^2 in the case of target ablation with 266 nm. The film surfaces produced with 266 nm wavelength show numerous droplets and high roughness (70 nm) (**Figure 11a**). In contrast, the films obtained by using 193 nm laser wavelength are smooth, without droplets (**Figure 11b**) and roughness below 2 nm.

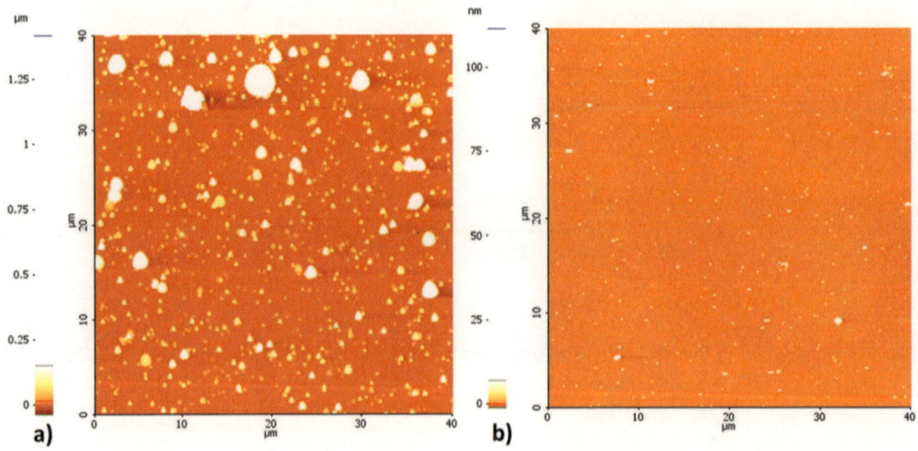

Figure 11. AFM images of Al_2O_3 thin films grown on Si substrates, at different wavelengths and laser fluences: (a) λ = 266 nm and Φ = 1.5 J/cm^2, and (b) λ = 193 nm and Φ = 6 J/cm^2.

From HRTEM investigation, polycrystalline Al_2O_3 thin layers are detected. Strong diffraction peaks corresponding to (110) silicon and weak diffraction peaks attributed to cubic Al_2O_3 (γ-Al_2O_3) structure (JCPDS file 29-0063) (space group SG:Fd3m and lattice constant a = 0.7924 nm) are shown in **Figure 12a**. The morphology of Al_2O_3 thin layers is shown in the HRTEM images at high magnification in **Figure 12b**. The layer is formed from small nanocrystallites (<10 nm).

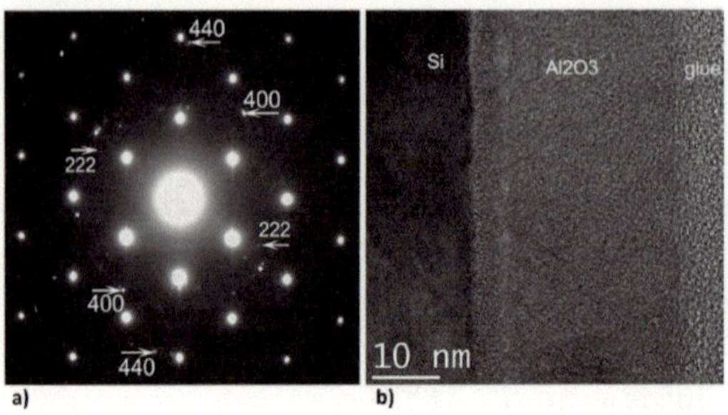

Figure 12. (a) Electron diffraction pattern of Al_2O_3 grown on Si substrate; (b) Cross-section HRTEM image of the Al_2O_3 thin film.

The optical properties of alumina thin films are studied by spectroellipsometry. Using 193 nm laser wavelength, thin films with refractive indices close to literature values are reported [35]. The thin layers grown from a target irradiated at 266 nm have low refractive indices ($n \sim 1.65$) due to the high presence of voids (**Figure 13**).

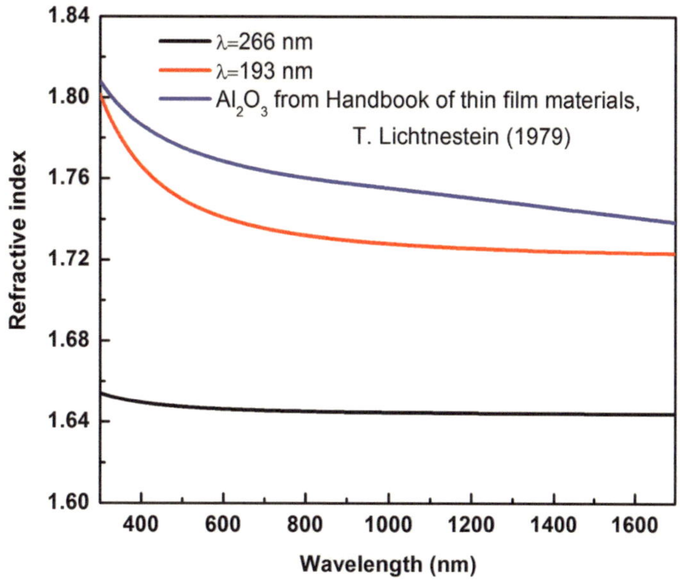

Figure 13. Refractive index of the Al_2O_3 thin films deposited by PLD compared to a reference Al_2O_3 thin film.

8. Conclusions

This chapter succinctly reviews the deposition of Si, Ta, Hf, and Al oxide materials as thin films by PLD and RF-PLD. All the results presented in this work are discussed in terms of two main applications, i.e. high-k dielectrics and antireflective coating applications. These applications have specific requirements, for example, the deposition of smooth and droplet-free thin films. This could be achieved by optimizing the laser wavelength, i.e. smooth and droplet-free Al_2O_3 could be obtained by using 193 nm laser wavelength. In contrast, smooth SiO_2 and HfO_2 thin films thin films with low roughness were obtained by 266 nm laser irradiation. In the case of Ta_2O_5, the thin films grown by RF-PLD at 266 nm wavelength have low roughness and in addition they present low leakage currents densities with three orders of magnitude smaller than for samples grown by PLD. Furthermore, the optical properties of the as-deposited metal-oxide thin films were investigated by spectroellipsometry, and, for example, the Ta_2O_5 thin films deposited by RF-PLD have a high refractive index (2.27) which makes them useful as antireflective coatings and waveguide material. In addition, we have also shown that hafnia thin films grown in low oxygen content ($\leq$ 0.5 mbar) have low leakage currents which make them suitable for applications in MOSFET devices.

To sum up, we have shown that by varying the deposition parameters in the classical and hybrid (RF-assisted) PLD process, it is possible to tune the morphology, structure, and optical properties of the Si, Hf, Ta, and Al oxide thin films to match antireflective coating and high-k dielectric applications.

Acknowledgements

The authors would like to acknowledge Dr. Valentin Ion from INFLPR for carrying out the ellipsometry measurements and Dr. Leona C. Nistor from the National Institute for Materials for providing the TEM images. Financial support from the Romanian National Nucleu Program — contract 4N/2016, the Romanian National Authority for Scientific Research, CNCS – UEFISCDI project number PN-II-PT-PCCA-2013-4-1870 (ARCOLAS), and PN-II-PT-PCCA-2013-4-1992 (SOLE) is gratefully acknowledged.

Author details

Mihaela Filipescu*, Alexandra Palla Papavlu and Maria Dinescu

*Address all correspondence to: mihaela.filipescu@gmail.com

National Institute for Lasers, Plasma and Radiation Physics, Magurele, Ilfov, Romania

References

[1] Ramanathan, S.; editor. Thin Film Metal-Oxides. 1st ed. New York: Springer US; 2010. 343 p. DOI: 10.1007/978-1-4419-0664-9

[2] Ratzke, M.; Wolfframm, D.; Kappa, M.; Koutev-Arguirova, S.; Reif, J.; Pulsed laser deposition of HfO and Pr_xO_y high-k films on Si(100). Appl Surf Sci. 2005;247:128-133. DOI: 10.1016/j.apsusc.2005.01.094

[3] Anghinolfi, L.; Prato, M.; Chtanov, A.; Gross, M.; Chincarini, A.; Neri, M.; Gemme, G.; Canepa, M.; Optical properties of uniform, porous, amorphous Ta_2O_5 coatings on silica: temperature effects. J. Phys. D: Appl. Phys. 2013;46:455301. DOI: 10.1088/0022-3727/46/45/455301

[4] Choi, G. M.; Tuller, H. L.; Haggerty, J. S.; Alpha - Ta_2O_5: an intrinsic fast oxygen ion conductor. J. Electrochem. Soc. 1989;136(3):835-838. DOI: 10.1149/1.2096752

[5] Ushikubo, T.; Recent topics of research and development of catalysis by niobium and tantalum oxides. Catal. Today. 2000;57(3-4):331-388. DOI: 10.1016/S0920-5861(99)00344-2

[6] Bang, S.B.; Chung, T.H.; Kim, Y.; Plasma enhanced chemical vapor deposition of silicon oxide films using TMOS/O2 gas and plasma diagnostics. Thin Solid Films. 2003;444(1-2):125-131. DOI: 10.1016/S0040-6090(03)01127-1

[7] Van Mol, A.M.B.; Chae, Y.; McDaniel, A.H.; Allendorf, M.D.; Chemical vapor deposition of tin oxide: fundamentals and applications. Thin Solid Films. 2006;502(1-2):72-78. DOI: 10.1016/j.tsf.2005.07.247

[8] Bae, D.; Kwon, S.; Oh, J.; Kim, W. K.; Park, H.; Investigation of Al_2O_3 diffusion barrier layer fabricated by atomic layer deposition for flexible Cu(In,Ga)Se_2 solar cells. Renew Energ. 2013;55:62-68. DOI: 10.1016/j.renene.2012.12.024

[9] Tomaszewski, H.; Haemers, J.; Denul, J.; De Roo, N.; De Gryse, R.; Yttria-stabilized zirconia thin films grown by reactive r.f. magnetron sputtering. Thin Solid Films. 1996;287(1–2):104-109. DOI: 10.1016/S0040-6090(96)08743-3

[10] Dinescu G.; Matei, D.; Brodoceanu, D.; Scarisoreanu, N.; Morar, M.; Verardi, P.; Craciun, F.; Toma, O.; Pedarnig, J.D.; Dinescu, M.; Influence of the radiofrequency plasma beam addition on the properties of pulsed laser deposited films. Proceedings SPIE. 2004;5448:136-143. DOI: 10.1117/12.548256

[11] Fujiwara. H.; Spectroscopic Ellipsometry Principles and Applications. 1st ed. Tokyo, Japan: Maruzen Co. Ltd; 2007.

[12] Herzinger, C.; Johs, B.; McGahan, W.A.; Woollam, J.A.; Paulson, W.; Ellipsometric determination of optical constants for silicon and thermally grown silicon dioxide via a multi-sample, multi-wavelength, multi-angle investigation. Appl. Phys. 1998;83:3323–3336.

[13] Hori T.; Gate Dielectrics and MOS ULSIs. 1st ed. Berlin Heidelberg: Springer; 1997. 343 p. DOI: 10.1007/978-3-642-60856-8

[14] Engelhart,P.; Hermann, S.; Neubert, T.; Plagwitz, H.; Grischke, R.; Meyer, R.; Klug, U.; Schoonderbeek, A.; Stute, U.; Brendel, R.; Laser ablation of SiO_2 for locally contacted Si solar cells with ultra-short pulses. Prog. Photovoltaics: Res. Applic. 2007;15:521-527.

[15] Millon, E.; Advanced functional oxide thin film grown by pulsed-laser deposition. Appl. Surface Sci. 2013;278:2-6. DOI: 10.1016/j.apsusc.2012.10.190

[16] Wilk, G.D.; Wallance, R.M.; Anthony, J.M.; High-k gate dielectrics: Current status and materials properties considerations. J. Appl. Phys. 2001;89(10):5243-5275. DOI: 10.1063/1.1361065

[17] Filipescu, M.; Ion, V.; Somacescu, S.; Mitu, B.; Dinescu, M.; Investigation of Ta_2O_5 and $TaSi_xO_y$ thin films obtained by radio frequency plasma assisted laser ablation for gate

dielectric applications. Appl. Surface Sci. 2013;276:691-696. DOI: 10.1016/j.apsusc. 2013.03.154

[18] Stanciu, G.; Filipescu, M.; Ion, V.; Dinescu, M.; Andronescu, E.; Optical properties of Ta_xO_y thin films obtained by laser deposition techniques. U.P.B. Sci. Bull. 2013;75(2): 15-22.

[19] Xiliang H.; Jiehua W.; Lili Z.; Jia M.; Xiangdong G.; Xiaomin L.; Synthesis and optical properties of tantalum oxide films prepared by ionized plasma-assisted pulsed laser deposition. Solid State Commun. 2008;147:90-93. DOI: 10.1016/j.ssc.2008.05.007

[20] Atanassova, E.; Spassov, D.; Paskeleva, A.; Influence of the metal electrode on the characteristics of thermal Ta_2O_5 capacitors. Microelectron. Eng. 2006;83:1918-1926. DOI: 10.1016/j.mee.2006.01.043

[21] Lee, Y.H.; Kwak, J.C.; Gang, B.S.; Kim, H.C.; Choi, B.H.; Jeong, B.K.; Park, S.H.; Lee K.H.; Photo-induced atomic layer deposition of tantalum oxide thin films from $Ta(OC_2H_5)_5$ and O_2. J. Electrochem. Soc. 2004;151(1):52-55. DOI: 10.1149/1.1629096

[22] Wei, A.X.; Ge, Z.X.; Zhao, X.H.; Liu, J.; Zhao, Y.; Electrical and optical properties of tantalum oxide thin films prepared by reactive magnetron sputtering. J. Alloys Compounds. 2011;509(41):9758-9763. DOI: 10.1016/j.jallcom.2011.08.019

[23] Traylor-Kruschwitz, J.D.; Pawlewicz, W.T.; Optical and durability properties of infrared transmitting thin films. Appl. Opt. 1997;36(10):2157-2159. DOI: 10.1364/AO. 36.002157

[24] Tu, Y.K.; Lin, C.C.; Wang, W.S.; Huang, S.L.; Optoelectronic materials devices packaging and interconnects. In: E. Batchman; R.F. Carson; R.L. Galawa; H.J. Wojtunik; editor. SPIE Proceedings. 836th ed. Bellingham, WA: SPIE; 1987. p. 40.

[25] Huang, A.P.; Yang, Z.C.; Chu, P.K.; Hafnium –based high-k gate dielectrics. In: P. K. Chu, editor. Advances in Solid State Circuit Technologies. 1st ed. Croatia: InTech; 2010. pp. 333-350.

[26] Lehan, J.P.; Mao, Y.; Bovard, B.G.; Macleod, H.A.; Optical and microstructural properties of hafnium dioxide thin films. Thin Solid Films. 1991;203(2):227-250. DOI: 10.1016/0040-6090(91)90131-G

[27] Baumeister P.; Arnon O; Use of hafnium dioxide in multilayer dielectric reflectors for the near UV. Appl. Opt. 1977;16(2):439-444. DOI: 10.1364/AO.16.000439

[28] Filipescu, M.; Scarisoreanu, N.; Craciun, V.; Mitu, B.; Purice, A.; Moldovan, A.; Ion, V.; Toma, O.; Dinescu, M.; High-k dielectric oxides obtained by PLD as solution for gates dielectric in MOS devices. Appl. Surf. Sci. 2007;253:8184-8191. doi:10.1016/j.apsusc. 2007.02.166

[29] Robertson, J.; Band offsets of wide-band-gap oxides and implications for future electronic devices. J. Vac. Sci. Technol. 2000; B(18):1785. doi:10.1116/1.591472

[30] Peacock, P. W.; Robertson, J.; Band offsets and Schottky barrier heights of high dielectric constant oxides. J. Appl. Phys. 2002; 92:4712-4721. doi:10.1063/1.1506388

[31] Riihelä, D.; Ritala, M.; Matero, R.; Leskelä, M.; Introducing atomic layer epitaxy for the deposition of optical thin films. Thin Solid Films. 1996:289(1-2): 250-255. doi:10.1016/S0040-6090(96)08890-6

[32] Smit, M.K.; Acket, G.A.; Van der Laan, C.J.; Al_2O_3 films for integrated optics. Thin Solid Films. 1986;138(2):171-181. doi:10.1016/0040-6090(86)90391-3

[33] Tauc. J.; Optical properties of amorphous semiconductors, J. Tauc (Eds), Amorphous and Liquid Semiconductors, Plenum Press, London and New York, 1974; 159-220. ISBN: 978-1-4615-8707-1

[34] Ion, M.; Berbecaru, C.; Iftimie, S.; Filipescu, M.; Dinescu, M.; Antohe, S.; PLD deposited Al_2O_3 thin films for transparent electronics. Digest J. Nanomater. Biostruct. 2012;7(4): 1609-1614. Accession Number: WOS:000312710300025

[35] Pillonnet,A.; Garapon, C.; Champeaux, C.; Bovier, C.; Brenier, R.; Jaffrezic, H.; Mugnier, J.; Influence of oxygen pressure on structural and optical properties of Al_2O_3 optical waveguides prepared by pulsed laser deposition. Appl. Phys. A. 1999;69:735-738. DOI: 10.1007/s003399900376

Chapter 4

Fluorinated Porphyrinic Crystalline Solids: Structural Elucidation and Study of Intermolecular Interactions

Subramaniam Sujatha and Chellaiah Arunkumar

Additional information is available at the end of the chapter

http://dx.doi.org/10.5772/62946

Abstract

Crystal engineering is an emerging area of research in material, biological, and pharmaceutical chemistry that involves synthesis of new materials, analysis of its structure including intermolecular interactions using X-ray crystallography as well as computational methods. It has been shown that the intermolecular interactions involving organic fluorine such as C–F•••H, F•••F, and C–F•••π play an important role in stabilizing the supramolecular assemblies, especially in the absence of strong intermolecular forces. Recently, non-covalent interactions involving conjugated aromatic system such as porphyrins have been studied intensively. The synthetic porphyrins are of widespread attention because of their close resemblance to naturally occurring tetrapyrrolic pigments and they find various materials and biological applications. In this book chapter, we disclose our recent findings on detailed crystal structure analysis of a few series of fluorinated porphyrins using single-crystal XRD as well as computational Hirshfeld surface analysis to understand the role of close contacts involving fluorine in the molecular crystal packing.

Keywords: Porphyrinoids, fluorinated porphyrins, crystal structure elucidation, intermolecular interactions, Hirshfeld surfaces

1. Introduction

Porphyrins are highly conjugated tetrapyrrolic pigments widely found in nature. They play a significant role in biological functions such as electron transfer, oxygen transport, and photosynthetic processes [1]. Most of the naturally occurring macro molecules such as hemoglobin chlorophyll, cytochromes have porphyrin as the basic unit. In these large classes of intensely colored macrocyclic pigments, the four pyrrole subunits are interconnected by methine bridges.

The synthesis of porphyrins can be achieved using the synthetic routes namely Rothemund method, Adler-Longo method, or Lindsey method [2–6]. Synthetic porphyrins are of two types, which include β-substituted and *meso*-substituted porphyrins. The structure of β-substituted porphyrins shows close resemblance to naturally occurring porphyrins, whereas the *meso*-substituted porphyrins display potential applications such as biomimetic photosynthesis, molecular electronics, supramolecular catalysis, and organic synthesis [7, 8]. Due to their promising photophysical properties, such as long wavelength absorption and emission, easy functionalization, high singlet oxygen quantum yield and low in vivo toxicity, they also found employed in magnetic resonance imaging (MRI) and photodynamic therapy (PDT) [9–11]. They are the fascinating molecules because of their distinct structure of altering the central metal atom, high thermal stability, tunable shape, symmetry, and synthetic versatility for their physico-chemical properties [8]. The position of the functional groups on the porphyrin building blocks alter the molecular packing and hence their supramolecular assemblies.

In recent years, fluorinated porphyrins are shown much attention since the pharmacological properties can be enhanced by the incorporation of fluorine atoms, and they are very effective in bio-medical applications [12, 13]. The reason behind the fact is that the little atom fluorine is having unique properties such as metabolic stability, binding selectivity, absorption, distribution, and excretion characteristics. Owing to the high electronegativity with its small size makes fluorine an ideal candidate for the replacement of hydrogen. Importantly, by increasing the lipophilicity and also by reducing the charges present in the molecule, fluorinated drugs possess excellent membrane penetration capability compared to its non-fluorinated counter parts [14]. The fluorinated drugs are numerous, and hence, the study of weak intermolecular interactions involving fluorine is essential in the field of medicinal chemistry to promote the discovery of new drugs of desirable profiles.

Intermolecular interactions are mainly of two categories [15], namely isotropic or directional (C•••C, C•••H, H•••H interactions) that define the size, shape as well as close packing and anisotropic or non-directional (hydrogen bonds, charge transfer interactions, halogen interaction, and heteroatom interactions). The weak individual intermolecular interactions act co-operatively and provide a better stability to the crystal packing. The analysis of nature and strength of intermolecular interactions is really important in the area of crystal engineering in order to design new materials especially drugs of desirable properties. They can be studied using X-ray crystallography as well as computational methods [16]. Interactions involving fluorine are of mainly three kinds, namely C–F•••H, F•••F, and C–F•••π which provides stability to form molecular self-assemblies especially in the absence of strong intermolecular forces.

Study of weak interactions involving porphyrins offers an interesting field of research in crystal engineering [17]. The energetically weak interactions such as van der Waals forces, hydrogen bonding, or metal coordination are the holding forces to make the self-assembly of porphyrins and quantifying these non-covalent interactions often brings a better understanding in these supramolecular self-assemblies. In this context, Hirshfeld surface (HS) analysis [18] is an important technique that helps to get the detailed information about the crystal structures in terms of intermolecular interactions. HSs and 2D fingerprint plots (FPs) were

http://dx.doi.org/10.5772/62946

generated using the single crystal X-ray diffraction data by Crystal Explorer 3.1 [19]. The term d_{norm} is a ratio of the distances of any surface point to the nearest interior (d_i) and exterior (d_e) atom and the van der Waals radii of the atoms [20]. The red color in the HSs indicates the closest contact bearing a negative value of d_{norm} with $d_i + d_e$ is shorter than the sum of the relevant van der Waals radii. Whereas the white color denotes the intermolecular distances close to van der Waals contacts with d_{norm} equal to zero. The contacts longer than the sum of van der Waals radii with positive d_{norm} values are blue in color. The 2D FP (d_i versus d_e) recognizes the presence of various non-covalent interactions present in the molecular crystals [21]. In this book chapter, we wish to discuss our recent observations on the detailed crystal structure analysis of a few series of fluorinated porphyrins (**Figure 1**) using single crystal X-ray crystallography. The role of weak intermolecular interactions in the molecular crystal packing was quantitatively analyzed using computational HS analysis by Crystal Explorer 3.1. Also, the role of organic fluorine towards crystal packing is discussed in detail.

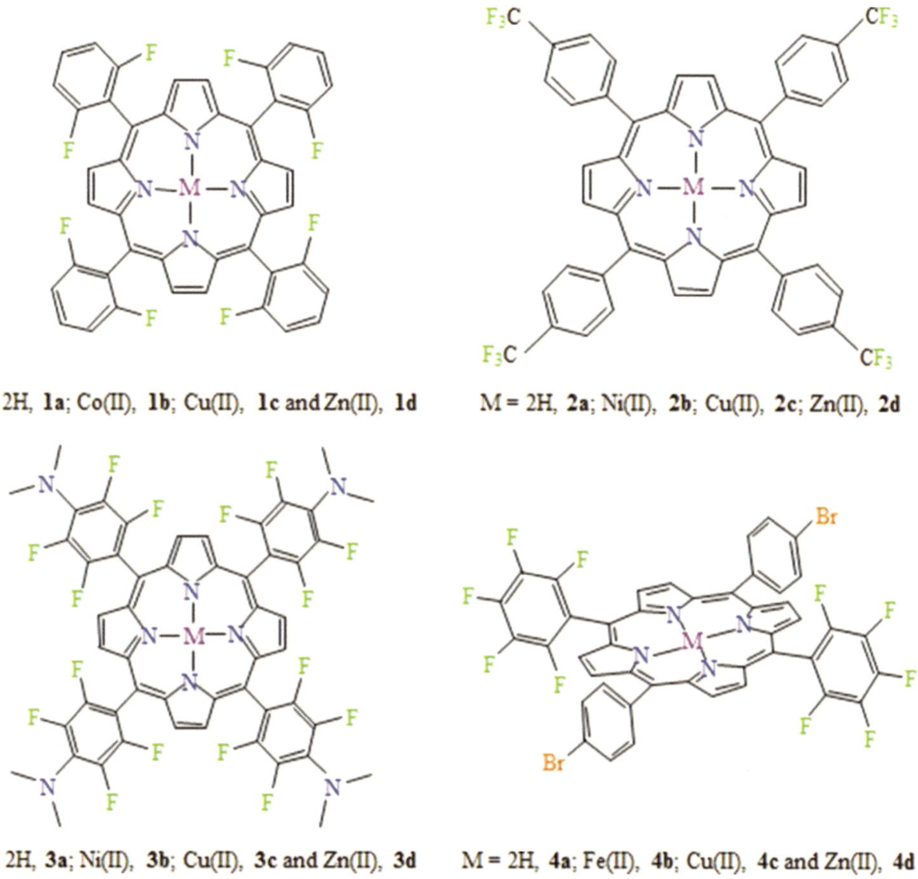

Figure 1. Chemical structures of fluorinated porphyrinic crystalline solids under study.

2. Synthesis of fluorinated porphyrins

Synthesis of 5,10,15,20-tetrakis(2′,6′-difluorophenyl)porphyrin, $H_2T(2',6'\text{-}DFP)P$, **1a** and 5,10,15,20-tetrakis-(4′-trifluoromethylphenyl)porphyrin, $H_2T(4'\text{-}CF_3P)P$, **2a**, were achieved by the treatment of pyrrole with 2,6-difluorobenzaldehyde/4-trifluoromethylbenzaldehyde using the method reported by Lindsey et al. [6]. The porphyrin, H_2TF_4DMAP (5,10,15,20-tetrakis(2′, 3′,5′,6′-tetrafluoro-*N,N*-dimethyl-4-anilinyl)porphyrin), **3a** was prepared using H_2TF_5PP (5,10,15,20-tetrakis(pentafluorophenyl)porphyrin) and dimethylamine hydrochloride in DMF at 180 °C under N_2 atmosphere using the modified procedure in 85% yield. Consuming 5-(pentafluorophenyl)dipyrromethane and 4-bromobenzaldehyde, the desired *trans*-porphyrin, 5,15-di(pentafluorophenyl)-10,20-bis(4′-bromophenyl)porphyrin was synthesized using the modified procedure of Lindsey et al. [22]. The metal complexes (**1b–1d, 2b–2d, 3b–3d,** and **4b–4d**) were prepared by the literature method [23] to afford the desired complexes in quantitative yields. The zinc(II) and copper(II) complexes were obtained in chloroform/methanol mixture using the corresponding metal acetates as metal ion carrier, whereas iron(II), cobalt(II), and nickel(II) complexes were prepared in DMF using metal chloride. All the synthesized porphyrins were isolated, purified by column chromatography and characterized by the conventional spectroscopic methods involving UV-Visible, ^1H NMR, mass spectrometry, and single-crystal X-ray diffraction analysis.

3. Structural determination of fluorinated porphyrins

The single crystals of the porphyrins were grown at room temperature by vapor diffusion method using appropriate solvents. The X-ray data were obtained using a Bruker AXS Kappa Apex II CCD diffractometer with graphite monochromated Mo Kα radiation ($\lambda = 0.71073$ Å) at room temperature. The structure solution was done using SIR92 [24] (WINGX32) program by direct methods, and the refinement (full-matrix least squares) was performed on $|F|^2$ using the SHELXL97 [25] software by successive Fourier synthesis with $I > 2\sigma(I)$ reflections.

3.1. Structural description of MT(2′,6′-DFP)P (1a–1d)

The porphyrin ligand **1a** containing difluorophenyl group and their metal complexes (**1b–1d**) are crystallized in monoclinic system with half a molecule in the asymmetric unit with space groups *P2₁/c, P2₁/n, P2₁/c,* and *P2₁/n C2/c,* respectively (**Table 1**). The ORTEP and crystal packing diagrams of **1b–1d** are shown in **Figure 2**. The compounds **1a, 1c,** and **1b, 1d** are isostructural with similar crystal packing pattern and the former pair crystallized without solvent molecules in the crystal lattice. The porphyrin plane of **1a** is not ideally planar with maximum deviation of 0.081(1) Å from the mean plane for N2 atom alone. The two difluorophenyl moieties, "C11–C16" and "C17–C22" are inclined to each other through a dihedral angle of 82.71°(6) and 65.85°(5). The distance of 2.686 Å from the centroid of the ring (N2•C6•C7•C8•C9) to H19 #1 is connected by c-glide (*Symm #1.* x, ½–y, -½+z) through C–H•••π interactions. Moreover, a 2D sheet of zigzag motifs of molecules formed through C–H•••F (C20–H20•••F3) and C–H•••π interactions. These 2D sheets are linked to produce the 3D network arrangement which is stabilized by the C–H•••F (C15–H15•••F1) interaction. The tetra coordinated copper atom in **1c** lies in the porphyrin core and exhibits perfect square planar geometry with N1–Cu1–N2 bit angle of 90° which makes the porphyrin moiety more planar than the free ligand, **1a**. The porphyrin macrocycle of **1b** is close to planar, the cobalt(II) ion exhibit octahedral geometry and the two methanol molecules coordinated at the apex position. The Co–N distances are equal and in agreement with the reported values [26]. The hydrogen of the coordinated methanol forms O–H•••π interactions with the phenyl ring "C17–C22", the shortest interaction distance of 2.485 (5) Å is given by O1–H1•••C19 #1 contact (Symm: #1 ½–x, ½+y, ½–z). In **1d**, the porphyrin core is planar and the zinc(II) ion is hexa-coordinated and stays in the plane of porphyrin moiety. The average Zn–N and Zn–O distances are 2.054(5) Å and 2.324(5) respectively which is well comparable with that of reported $ZnTPP(THF)_2$ complex [27].

There are no direct π–π and intramolecular interactions seen from the geometrical analysis on the crystal structures of porphyrins, **1a–1d**. But, the non-covalent interactions like, $_{(ph,sol)}$C–H•••F, C–H•••$\pi_{(ph,pyr)}$, O–H•••$\pi_{(ph,pyr)}$, C–F•••$\pi_{(ph,sol)}$, and C–H•••O are present in the molecular crystal packing. In addition to these, C–H•••N and F•••F interactions were also present in **1a** and **1b, 1d,** respectively. The F•••F contact distance in **1b** and **1d** is found to be 2.901 and 2.782 Å, respectively. The crystal structures of **1a** and **1c** mainly contains three types of interactions in the crystal packing such as $_{(ph)}$C–H•••F, C–H•••$\pi_{(pyr)}$, C–F•••$\pi_{(ph)}$, and their distances are in the range of 2.471–2.652 Å, 2.760–2.766 Å and 3.063–3.165 Å, respectively. While in **1b** and **1d**, the types are $_{(ph,sol)}$C–H•••F, C–H•••$\pi_{(ph,pyr)}$, O–H•••$\pi_{(ph,pyr)}$, F•••F, and C–H•••O interactions.

	1a	**1b**	**1c**	**1d**
Empirical formula	$C_{44}H_{22}F_8N_4$	$C_{46}H_{28}CoF_8N_4O_2$	$C_{44}H_{20}CuF_8N_4$	$C_{46}H_{28}ZnF_8N_4O_2$
Fw	758.66	879.65	820.18	886.09
Color	Purple	Pink	Red	Red
Crystal system	Monoclinic	Monoclinic	Monoclinic	Monoclinic
Space group	P21/c	P21/n	P21/c	P21/n
a, Å	12.5342(10)	12.426(5)	12.5477(7)	12.4217(18)
b, Å	11.4384(9)	12.723(5)	11.3955(6)	12.529(2)
c, Å	12.1720(10)	12.709(5)	12.1642(7)	12.6954(18)
α, (°)	90.0	90.000(5)	90.0	90.0
β, (°)	97.080(3)	114.615(5)	96.391	114.092(17)
γ, (°)	90.0	90.000(5)	90.0	90.0
vol (Å^3)	1731.8(2)	1826.7(13)	1728.52(17)	1803.7(5)
Z	2	2	2	2

	1a	1b	1c	1d
D_{calcd} (mg/m^3)	1.455	1.599	1.576	1.632
Wavelength (λ), Å	0.71073	0.71073	0.71073	0.71073
T (K)	293(2)	293(2)	293(2)	293(2)
No. of unique reflections	2690	3583	5329	3165
No. of parameters refined	258	278	259	351
GOF on F^2	1.026	1.055	1.013	1.059
R_1^{a}	0.0359	0.0488	0.0372	0.0766
wR_2^{b}	0.0818	0.1445	0.0944	0.1685

$^aR_1 = \Sigma||F_o|-|Fc||/\Sigma|F_o|; I_o > 2\sigma (I_o)$.

$^bwR_2 = [\Sigma\ w(F_o^2-F_c^2)^2/\Sigma w(F_o^2)^2]^{1/2}$.

Table 1. Crystal structure data of porphyrins under study, **1a**, H_2T(2′,6′-DFP)P; **1b**, CoT(2′,6′-DFP)P. 2(MeOH); **1c**, CuT(2′,6′-DFP)P; **1d**, ZnT(2′,6′-DFP)P. 2(MeOH).

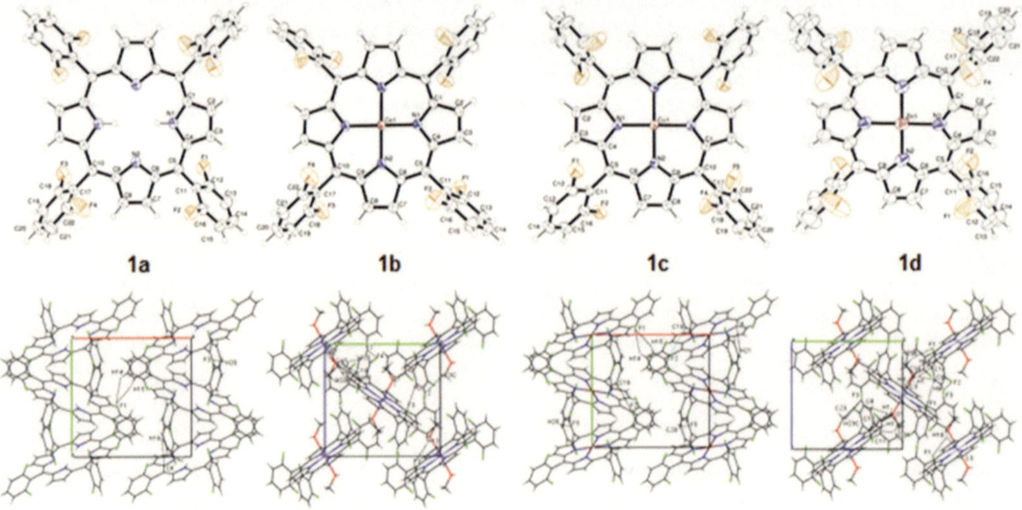

Figure 2. ORTEP and molecular crystal packing diagrams of **1a–1d**.

In order to quantify the various intermolecular interactions, HSs and their associated finger print plots were calculated using *Crystal Explorer 3.1*. The HSs of porphyrins, **1a–1d** are illustrated in **Figure 3** showing surfaces that have been mapped with d_{norm}. The molecular HSs generated for the two isostructural pairs **1a, 1c,** and **1b, 1d** have almost similar shapes reflecting the similar kind of crystal packing modes. The crystal packing of **1b** is mainly controlled by dominant interactions between methanolic OH and phenyl carbon atom whereas in **1d**, between methanolic OH with pyrrole carbon atom observed as strong red spots. The medium intense red spots are observed for F•••$H_{(ph)}$ contacts in **1a**, **1c** and $_{(pyr)}$H•••$O_{(MeOH)}$ contacts in **1b**, **1d**.

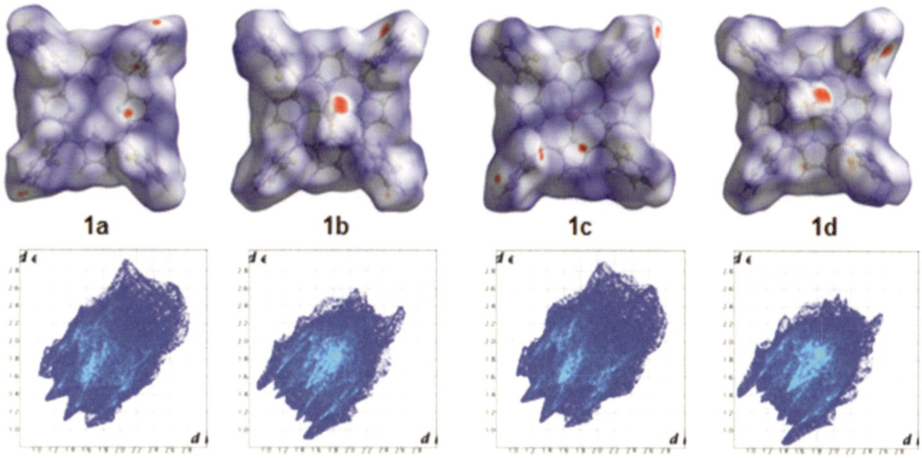

Figure 3. Hirshfeld surfaces mapped with d_{norm} ranging from -0.40 Å (red) to 2.50 Å (blue); 2D finger print plots with d_i and d_e ranging from 1.0 to 2.8 Å for **1a–1d.**

From the FP plots, the division of contributions is possible for different interactions including C•••H, F•••H/C/F, H•••H, and N•••H through interactive computer graphics which commonly overlap in the full FP plots. The shapes of FP plots of the two isostructural pairs are comparable The FPs of **1a–1d** features spike of various lengths and thickness, and the most prominent being the presence of wing-like peripheral spikes for C•••H contact at the top left and bottom right of each plot and a similar feature is also observed in small organic molecules [18, 28]. In **1a–1d**, 26–28% of intermolecular contacts are associated with C•••H and the distance ($d_i + d_e$) is about 2.65 Å for **1a**, **1c** and 2 Å for **1b**, **1d**. The H•••F contact is characterized by a pair of small spikes which also accounts 27–28% of the HSs in **1a–1d**. The van der Waals radii of fluorine atom is found to be in the range of 1.35–1.50 Å [29, 30]. The organic fluorine's role in crystal packing of **1a–1d** is so prominent because of the F•••H contact distances observed are significantly shorter than the corresponding sum of van der Walls radii (2.45–2.70 Å). The presence of broader spikes (24–34%) for H•••H contacts are in the range of 2.52–2.70 Å [31] which further strengthens the crystal lattice in **1a–1d**. The interactions involving fluorine other than H•••F contacts are F•••F (4–6% in **1a–1d**), characterized by broad peak; C•••F (7% in **1a**, **1c**; 2% in **1b**, **1d**), characterized by small curved lines/small peaks. The C•••F and F•••F contact distances ($d_i + d_e$) are seen in the range of 3.28–3.48 Å and 2.80–3.02 Å, respectively, which are nearer to the sum of its van der Walls radii (3.03–3.27 Å and 2.70–3.00 Å). Overall, the weak interactions involving fluorine is 35–40%, they can play a crucial role in directing the supramolecular organization of porphyrins [32]. **Figure 4** shows the distribution of individual intermolecular interactions on the basis of HS analysis for the related compounds.

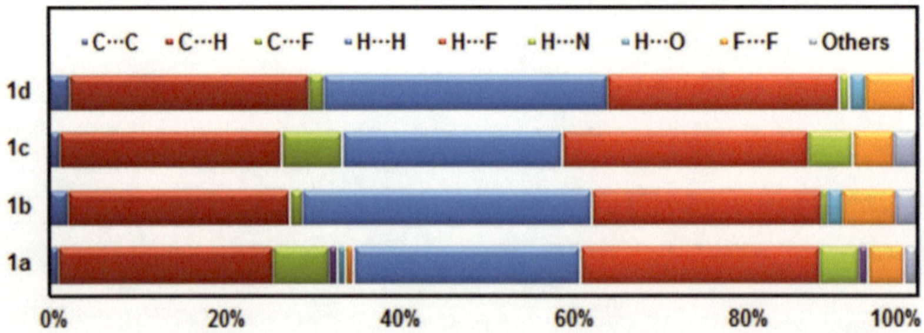

Figure 4. Distribution of individual intermolecular interactions on the basis of Hirshfeld surface analysis of **1a–1d.**

3.2. Structural description of MT(4′-CF$_3$P)P (2a–2d)

Compounds, **2a** and **2c** were crystallized in triclinic, whereas **2b** and **2d** in monoclinic system (**Table 2**). In all the porphyrins, the solvent molecule THF is either present in the crystalline lattice or they bind to the metal centre which show the well-known envelope conformation as witnessed in the literature [33, 34]. The asymmetric unit of **2a** consists of two half molecules of porphyrins and two THF molecules. The porphyrin core in **2a** exhibit planar skeleton, whereas in **2c,** it is of non-planar, possibly due to the presence of 4-trifuloromethylphenyl groups and an extra THF molecule which makes them being non-isostructural. In compound **2b**, the asymmetric unit contains half molecule of porphyrin and one THF molecule is coordinated to the nickel(II) centre. The ORTEP diagram of **2b** consists nickel(II) ion which is hexa-coordinated with two THF molecules at the apex positions and the molecular crystal packing is formed through C–F•••$H_{(sol)}$ interactions (**Figure 5**). The bond angles N–Ni–$N_{(adj)}$ and N–Ni–$N_{(opp)}$ in **2b** indicates a perfect planar geometry of the porphyrin core.

	2a	2b	2c	2d
Empirical formula	$C_{56}H_{42}F_{12}N_4O_2$	$C_{56}H_{40}F_{12}N_4NiO_2$	$C_{52}H_{32}CuF_{12}N_4O$	$C_{52}H_{32}F_{12}N_4OZn$
Fw	1030.94	1087.63	1020.36	1022.19
CCDC no.	1013690	1013689	1011088	1011087
Colour	Purple	Violet	Brown	Brown
Crystal system	Triclinic	Monoclinic	Triclinic	Monoclinic
Space group	P-1	P2$_1$/n	P-1	P21/c
a, Å	11.5014(8)	14.5791(19)	9.7136(5)	19.427(3)
b, Å	13.1179(12)	9.5303(19)	14.5452(9)	9.3940(10)
c, Å	17.3269(15)	18.351(4)	16.8870(13)	25.466(5)
a, (°)	96.329(3)	90.0	76.608(3)	90.0

http://dx.doi.org/10.5772/62946

	2a	2b	2c	2d
β, (°)	92.833(3)	101.571(8)	89.617(3)	102.756(5)
γ, (°)	108.739(3)	90.0	76.676(3)	90.0
Volume (Å^3)	2450.5(4)	2498.0(8)	2255.8(3)	4532.8(12)
Z	2	2	2	4
D_{calcd}(mg/m^3)	1.397	1.446	1.502	1.498
λ, Å	0.71073	0.71073	0.71073	0.71073
T (K)	173	173	173	173
No. of unique reflections	8528	4413	7810	7986
No. of parameters refined	788	397	825	813
GOF on F^2	1.027	1.015	1.039	1.022
R_1[b]	0.0657	0.0477	0.0476	0.0465
wR_2[c]	0.1849	0.1179	0.1205	0.1129

[a] $R_1 = \Sigma||F_o|-|Fc||/\Sigma|F_o|$; $I_o > 2\sigma(I_o)$.

[b] $wR_2 = [\Sigma w(F_o^2-F_c^2)^2/\Sigma w(F_o^2)^2]^{1/2}$.

Table 2. Crystal structure data of porphyrins under study: 2a, $H_2T(4'\text{-}CF_3P)P$. $(THF)_2$; 2b, $Ni^{II}T(4'\text{-}CF_3P)P$. $(THF)_2$; 2c, $Cu^{II}T(4'\text{-}CF_3P)P$. (THF); 2d, $Zn^{II}T(4'\text{-}CF_3P)P$. (THF).

The copper(II) centre in **2c** is four coordinated with the inner core nitrogens of the porphyrin ring, exhibits non-planarity and the mean plane displacement of the core atoms is ±0.3313 Å

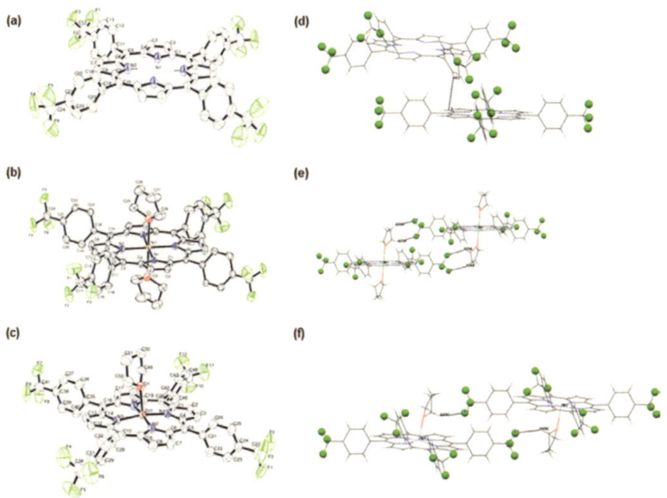

Figure 5. ORTEP diagrams and slip-stack orientations of molecules of **2a**, **2b** and **2d** (Solvent molecules and hydrogens are not shown for clarity, thermal ellipsoids are shown at 40% probability level) through C–H•••π; C–F•••$H_{(sol)}$ and C –F•••$H_{(sol)}$ interactions, respectively.

which is less compared to that of the reported β-pyrrole substituted copper(II) porphyrins [35, 36]. The ORTEP diagrams (top and side view) and packing diagram showing the slip-stack orientation of molecules through C–H•••π interactions are also shown in **Figure 6**. The non-planarity of the core is possibly due to 4-trifluoromethylphenyl groups and solvent (THF) molecule present in the lattice.

The zinc(II) centre in **2d** is five coordinated, which exhibits domed shape which is quite known for the zinc(II) complexes and the porphyrin plane is almost planar. However, the nitrogen atoms, N3 and N4 are above the plane with 0.078(3) Å and 0.045(3) Å; N1, N2, and Zn atoms are below the plane with 0.107(3) Å, 0.074(3) Å and 0.218(1) Å, respectively [37]. The bound THF is responsible for the metal to be slightly pulled down from the mean plane.

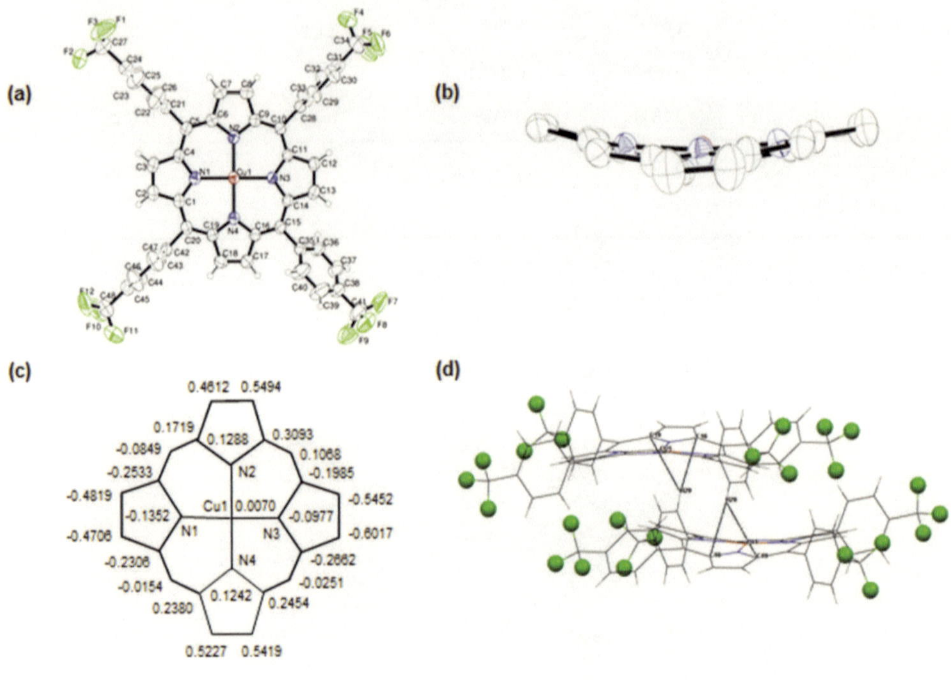

Figure 6. ORTEP diagrams of **2c**, (a) top view; (b) side view (solvent molecule is not shown for clarity; thermal ellipsoids are shown at 40% probability level); (c) mean plane displacements of the Cu(II) ion and the core atoms shown in Å units; (d) slip-stack orientation of molecules through C–H•••π interactions viewed down 'c' axis.

The geometrical analysis of porphyrins, **2a–2d** reveals that the molecular crystal packing consists of intermolecular interactions viz., $_{(sol/ph/pyr)}$C-H•••F, $_{(sol/ph/pyr)}$C-H•••C$_{(sol/ph/pyr)}$, C-F•••C$_{(ph/pyr)}$, $_{(ph/pyr)}$H•••H$_{(sol)}$, $_{(pyr)}$C–H•••O$_{(sol)}$, and F•••F. Interestingly, the H•••H close contact is seen only in **4** and **5**, and the distance varies from 1.845 to 2.364 Å which is considerably shorter compared with the previously discussed fluorinated porphyrins. The HSs, which have been mapped over a d_{norm} range of -0.14 to 2.0 Å, are illustrated in **Figure 7**.

http://dx.doi.org/10.5772/62946

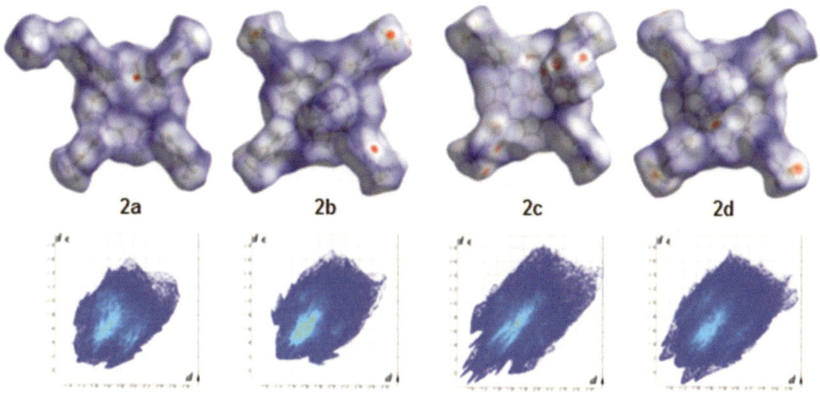

Figure 7. Hirshfeld surfaces of porphyrins, **2a–2d** with d_{norm} mapped ranging from -0.14 (blue) to 2.02 (red); 2D FPs for the various intermolecular interactions with d_i and d_e ranging from 2.31 to 2.44 Å.

The crystal packing in porphyrins **2a–2d** is mainly controlled by F•••H interaction that is observed as intense red spots on the HSs. The F•••F contacts are also seen as large intense red spots in **2a–2c**. The F•••C contacts are observed as medium intense red spots in free ligand **1a** and its copper derivative **2c** whereas faint red area in zinc derivative **2d**.

Apart from this, porphyrin **2d** shows faint red spots for N•••F close contact. In addition to the close contacts involving fluorine, C•••H contacts are appeared as intense and/or faint red spots in porphyrins **2a** and **2c** and the red spots are also seen in H•••H contacts of **2a** and **2b.** The shape of the FPs of porphyrins in **2a–2d** is unique indicating that there are no isostructural pair and the most dominant interaction is F•••H contacts seen at the middle region of each plot (**Figure 7**). In **3** and **4**, 44% of intermolecular contacts are associated with F•••H contacts, whereas in **2a** and **2b,** they are associated with 36 and 39% of the HSs respectively. And the distance ($d_i + d_e$) is about 2.31–2.44 Å for **2a–2d** which are significantly shorter than the corresponding sum of van der Walls radii (2.45–2.70 Å) indicating its significant role in crystal packing [29, 30]. The interactions involving fluorine other than F•••H contacts are F•••F and F•••C (4–5%) with $d_i + d_e$ value in the range of 2.59–2.96 Å and 3.14–3.20 Å, respectively. Also, the C•••H contacts are appeared at the top left and bottom right of each plot, and the percentage contributions are 12–18%, respectively, for **2a–2d**. The FPs of H•••H contact are somewhat closer in nature compared to the previously discussed fluorinated porphyrins. Overall, the interactions involving fluorine are 46% for **2a** and 53–55% for **2b–2d**, which plays a major role in directing the supramolecular architecture of porphyrins (**Figure 8**) [38]. Apart from those above, the presence of other contacts such as O•••H, N•••H and N•••F interactions were also observed.

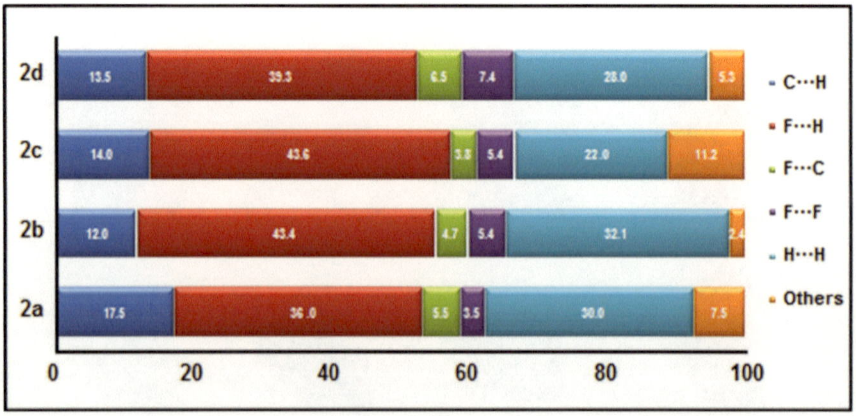

Figure 8. Percentage contribution of non-covalent interactions in porphyrins, **2a–2d** on the basis of HSs.

3.3. Structural description of MTF_4DMAP (3a–3d)

Porphyrins **3a** and **3c** are isostructural, crystallized in triclinic system (**Table 3**) and the asymmetric unit consists of crystallographically independent two halves of molecules with disordered water molecules in the crystal lattice (**Figure 9**). In **3a**, the porphyrin moiety of both molecules is planar within the limits of standard deviation and the conformation of the two molecules is slightly different. The packing is mostly stabilized through four, fairly strong, intermolecular C–H•••F hydrogen bonds. In **3c**, the copper(II) ion is located at an inversion centre with square planar environment. The Cu–N bond length in equatorial plane varies between 1.976(3) and 1.999(3) Å. The $(N–M–N)_{adj}$ angle is 90° and the $(N–M–N)_{opp}$ angle is 180° indicating the perfect square planar geometry around the Cu(II) metal centre.

	3a	3b	3c	3d
Empirical formula	$C_{52}H_{40}F_{16}N_8O_2$	$C_{60}H_{48}F_{16}N_8NiO_2$	$C_{52}H_{36}CuF_{16}N_8O_2$	$C_{64}H_{56}F_{16}N_8O_3Zn$
Fw	1112.92	1275.77	1172.43	1354.54
Color	Purple	Pink	Red	Purple
Crystal system	Triclinic	Tetragonal	Triclinic	Tetragonal
Space group	P-1	$P4_2/n$	P-1	I4
a, Å	10.2706(5)	15.9979(5)	10.2273(5)	16.4305(3)
b, Å	14.6333(7)	15.9979(5)	14.6303(6)	16.4305(3)
c, Å	18.0523(9)	10.9552(7)	17.9791(7)	11.1822(5)
α, (°)	86.258(2)	90	86.664(2)	90
β, (°)	86.484(2)	90	86.704(2)	90
γ, (°)	74.150(2)	90	74.494(2)	90
vol (Å^3)	2601.7(2)	2803.8(2)	2585.42(19)	3018.76(16)

	3a	3b	3c	3d
Z	2	2	2	2
D_{calcd} (mg/m^3)	1.421	1.511	1.506	1.490
wavelength (λ), Å	0.71073	0.71073	0.71073	0.71073
T (K)	293(2)	293(2)	293(2)	293(2)
No. of unique reflections	6143	2169	5964	2605
No. of parameters refined	798	197	779	221
GOF on F^2	1.111	1.075	1.103	1.077
R_1[a]	0.0640	0.0522	0.0428	0.0410
wR_2[b]	0.1930	0.1529	0.1151	0.1083

[a]$R_1 = \Sigma||F_o|-|Fc||/\Sigma|F_o|$; $I_o > 2\sigma(I_o)$.

[b]$wR_2 = [\Sigma\, w(F_o^2-F_c^2)^2/\Sigma w(F_o^2)^2]^{1/2}$.

Table 3. Crystallographic data of porphyrins, 3a–d.

Compound **3b** and **3c** crystallizes in tetragonal system with space group $P4_2/n$ and I4, respectively, and the metal centre is four coordinated. The porphyrin macrocycle in **3b** is non-planar, and there is a fairly strong intermolecular halogen–π interaction given by C3···F2 #1 (*Symm #1: 1−x, 1−y, z*) with distance 2.992(5) Å and another weak halogen–π intermolecular interaction by C10···F1 #1 with distance 3.152(6) Å. The bond angles, $(N–M–N)_{adj}$ and $(N–M–N)_{opp}$ are 90° and 173.0°, respectively, indicating that the Ni(II) centre in **3b** is deviated from square planar geometry. The tetra-coordinated porphyrins, **3b** and **3c** possess the average M–N bond lengths of 1.938(3) Å and 1.989(3) Å, respectively. Porphyrin **3d** is penta-coordinated with a THF molecule and the asymmetric unit consists of one-fourth of the porphyrin molecule solvated with half of the THF molecule in the crystal lattice.

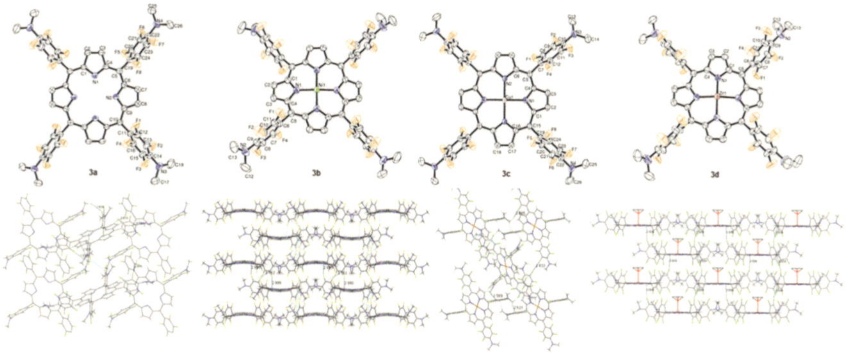

Figure 9. ORTEP and molecular crystal packing diagrams of **3a–3d.**

The average Zn–N and Zn–O distance in **3d** is 2.052(3) Å and 2.130(6) Å respectively and the values are in good agreement with that of the reported penta-coordinated Zn(II) complexes [24]. In **3a** and **3c,** there is a strong C···H interaction involving $C(NMe_2)$–H···$C_{(porp)}$ observed. Apart from these, several other intermolecular interactions involving fluorine such as C–H···F contact in which the fluorine interacts with the hydrogen of the *β*–pyrrole ring (2.338–2.799 Å) and C–F···π and F···F interactions are observed.

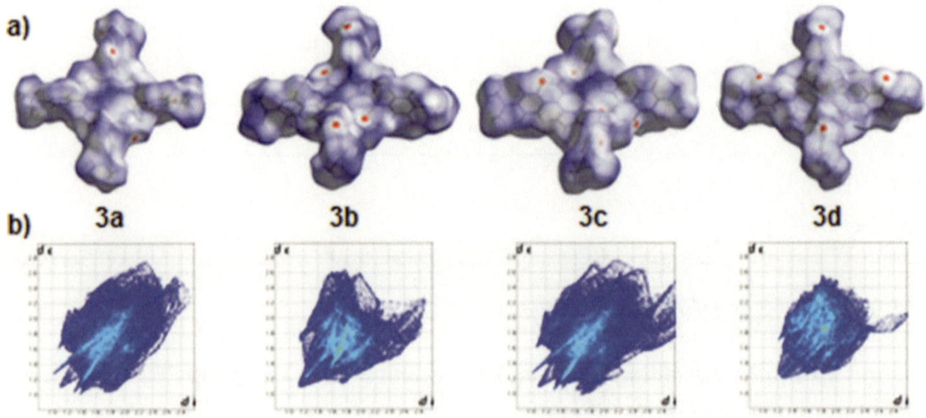

Figure 10. Hirshfeld surfaces with normalized contact distance ranging from -0.4 Å (red) to 2.0 Å (blue) and (b) 2D finger print plots of all the intermolecular contacts with d_i and d_e ranging from 1.0 to 2.8 Å for **3a–3d.**

The HSs of molecules **3a** and **3c** have similar shapes indicating similar interactions in the packing modes (**Figure 10**). The presence of the coordinated THF molecule to the Zn(II) is reflected in different size and shape of the HS at the centre of **3d** as compared to the flat surface at the centre of **3a–3c**. The crystal packing of **3a** and **3c** is mainly controlled by dominant interactions involving C•••H, F•••H, and F•••F contacts observed as red spots on the HSs whereas in **3b** it is F•••C and in **3d** it is of F•••H contacts. In the isostructural crystals, **3a** and **3c**, 15% of intermolecular contacts are associated with C•••H, whereas in **3b** and **3d** it is 12 and 26% of the HSs and the distance ($d_i + d_e$) is about 2.80 Å for **3a–3d**. The interactions involving fluorine are H•••F (43% in **3a**, **3c**; 33% in **3c**; 23% in **3d**); F•••F (6–7% in **3a–3d**). The $d_i + d_e$ value observed for H•••F contacts are about 2.40 Å for **3a**, **3c**; 2.70, and 1.85 Å for **3b**, **3d,** respectively, whereas F•••F contacts are in the range of 2.80–3.25 Å. In **3a** and **3c**, the different types of C–F···π interactions has the least contribution towards the total intermolecular interactions (3.0%), and the shortest close-contact distance ($d_i + d_e$) is about 3.2 Å. For the compounds **3b** and **3d**, it is found to be 11 and 16%, respectively, with $d_i + d_e$ is about 3.0 Å. As inferred from the geometrical analysis, the C–F···π interaction is primarily responsible for the crystal packing in **3b** indicated by the red spots in the HS (**Figure 10**). The H•••H contacts are characterized by sharper spikes in **3a–3d** (17–30%). **Figure 11** shows the contributions of various types of interactions in the HSs for all the investigated porphyrins. The decomposition of the FPs emphasizes that the weak interactions involving fluorine atom can act as the major

http://dx.doi.org/10.5772/62946

contributors for the crystal packing in addition to the C•••H contacts and the cooperativity of the H•••H contacts [39].

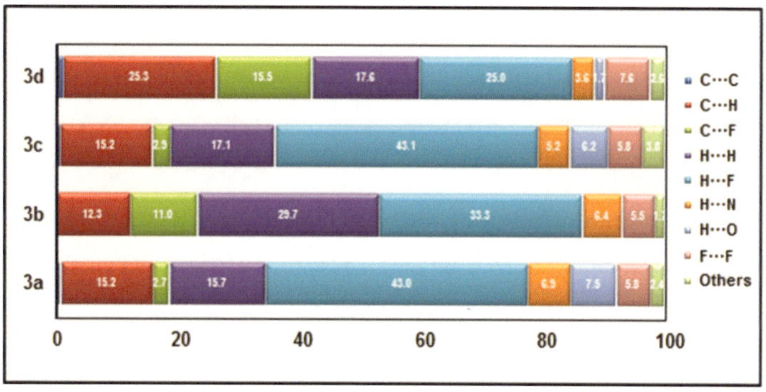

Figure 11. Relative percentage contribution of various intermolecular interactions in **3a–3d** based on Hirshfeld surface analysis.

3.4. Structural description of 5,15-di(pentafluorophenyl)-10,20-bis(4′-bromophenyl)porphyrin and its metal complexes (4a–4d)

Trans-porphyrins, **4a** and **4c** were crystallized in monoclinic with $P2_1/c$ space group whereas **4b** and **4d** in triclinic with P-1 space group (**Table 4**). The free ligand and the Cu(II) complex crystallized without solvent molecules. In **4b,** there are two pyridine molecules present at the apex positions and in **4d,** one of the THF molecule is coordinated to the zinc(II) centre, while another one is seen in the crystal lattice. As seen earlier, the bound or unbound THF solvate molecules in **4d** show the well-known envelope conformation [33, 34]. The asymmetric unit of **4a** and **4c** consists of half molecule of porphyrin in which both porphyrin cores are found to be planar in nature. The copper(II) centre in **4c** is tetra-coordinated with the inner core nitrogen of the porphyrin ring. The ORTEP and packing diagrams of the isostructural **4a** and **4c** are shown in **Figure 12**. The asymmetric unit of **4b** contains two half molecules of porphyrin each bearing a pyridine residue as solvent attached to the iron(II) centre. The iron(II) ion is hexa-coordinated with two pyridine molecules at the apex positions and the molecular crystal packing diagrams of **3** forms a orthogonal arrangements in a one dimensional array of molecules through C•••H and F•••F interactions viewed down 'c' axis. The zinc(II) centre in **4d** is penta-coordinated and exhibits domed shape which is quite common for the zinc(II)-porphyrin complexes. The porphyrin plane with the metal atom is almost planar. However, the atoms N1, N4, and Zn are above the plane with 0.0896(5) Å, 0.1081(5) Å, and 0.1409(1) Å; N2 and N3 atoms are below the plane with 0.3491(5) Å and 0.2406(5) Å, respectively.

Like other fluorinated porphyrins discussed earlier, the crystal packing of trans porphyrins **4a –4d** consists of a number of intermolecular interactions viz., $_{(sol/ph/pyr)}$C-H•••F, $_{(sol/ph/pyr)}$C-H•••Br, $_{(ph)}$C-F•••Br, $_{(sol/ph/pyr)}$C-H•••C$_{(sol/ph/pyr)}$, $_{(ph)}$C-F•••C$_{(ph)}$, $_{(ph/pyr)}$C•••C$_{(sol/ph)}$, $_{(ph/pyr)}$H•••H$_{(sol)}$, $_{(pyr)}$C–H•••O$_{(sol)}$ and F•••F. Interestingly, the symmetrical (F•••F) and unsymmetrical ($_{(ph)}$C-F•••Br) halogen interactions have been identified only in **4a** and **4c**, and the distance varies from 3.280 to 3.295 Å which is further supported by the HS analysis (**Figure 13**). As anticipated, the free ligand (**4a**) and its copper derivative (**4c**) are isostructural. The HSs mapped with d_{norm} for **4a** and **4c** highlights medium intense red spots for F•••H/F and C•••H close contacts and faint red spots for F•••C/Br contacts. The compound **4b** shows intense red spots for F•••H close contact and faint red spots for H•••Br contact and very pale red spots for F•••H contact, whereas **4d** shows several bright red spots for F•••H, few intense red spots for H•••Br close contact and faint red spots for N/C•••H and C•••F contact. Over all, porphyrin **4d** shows large number of red spots compared to other porphyrins **4a–4c** due to the role of lattice as well as coordinated THF molecule toward crystal packing.

	4a	**4b**	**4c**	**4d**
Empirical formula	$C_{44}H_{18}Br_2F_{10}N_4$	$C_{54}H_{26}Br_2F_{10}FeN_6$	$C_{44}H_{16}Br_2CuF_{10}N_4$	$C_{52}H_{32}Br_2F_{10}N_4O_2Zn$
Fw	952.44	1164.48	1013.97	1160.01
CCDC no.	1013690	1013689	1011088	1011087
Colour	Purple	Red	Purple	Purple
Crystal system	Monoclinic	Triclinic	Monoclinic	Triclinic
Space group	$P2_1/c$	P-1	$P2_1/c$	P-1
a, Å	14.9457(6)	10.700(5)	14.8461(9)	12.6848(5)
b, Å	8.9793(4)	13.933(5)	8.9921(5)	13.4661(5)
c, Å	15.9316(6)	17.329(5)	15.8111(9)	16.2952(6)
α, (°)	90	103.791(5)	90	66.724(2)
β, (°)	116.550(2)	100.709(5)	116.115(3)	68.750(2)
T (K)	293(2)	293(2)	293(2)	293(2)
λ, Å	0.71073	0.71073	0.71073	0.71073
γ, (°)	90	103.144(5)	90	76.468(2)
Volume (Å^3)	1912.58(14)	2363.2(16)	1895.26(19)	2369.64(16)
Z	2	2	2	2
D_{calcd} (mg/m^3)	1.654	1.636	1.777	1.626
No. of unique reflections	3747	6952	3340	8341
No. of parameters refined	277	661	277	854
GOF on F^2	1.022	1.057	1.025	1.024
R_1^b	0.0465	0.0551	0.0448	0.0586
wR_2^c	0.1317	0.1395	0.1131	0.1468

$^aR_1 = \Sigma||F_o|-|Fc||/\Sigma|F_o|; I_o > 2\sigma (I_o)$.

$^bwR_2 = [\Sigma\, w(F_o^2-F_c^2)^2/\Sigma w(F_o^2)^2]^{1/2}$.

Table 4. Crystal structure data of porphyrins under study, MBPFPBBPP; M = 2H, 4a; Fe(II). (pyridine)$_2$, 4b; Cu(II), 4c and Zn(II). (THF)$_2$, 4d.

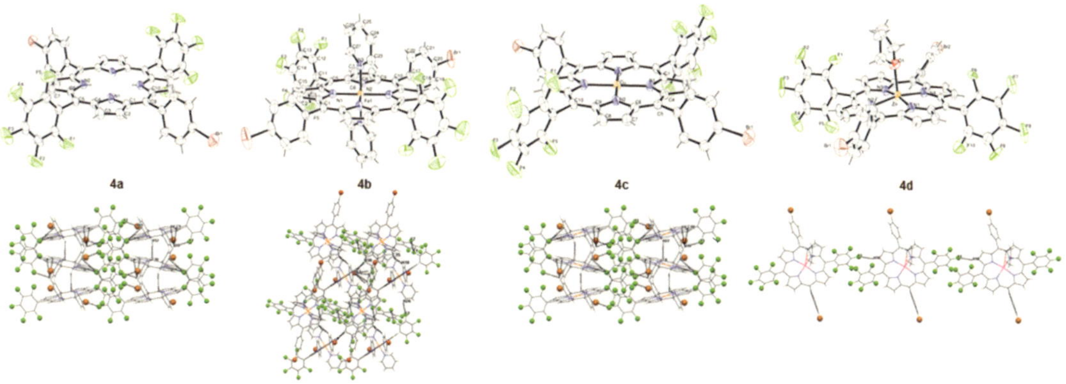

Figure 12. ORTEP and molecular crystal packing diagrams of **4a–4d.**

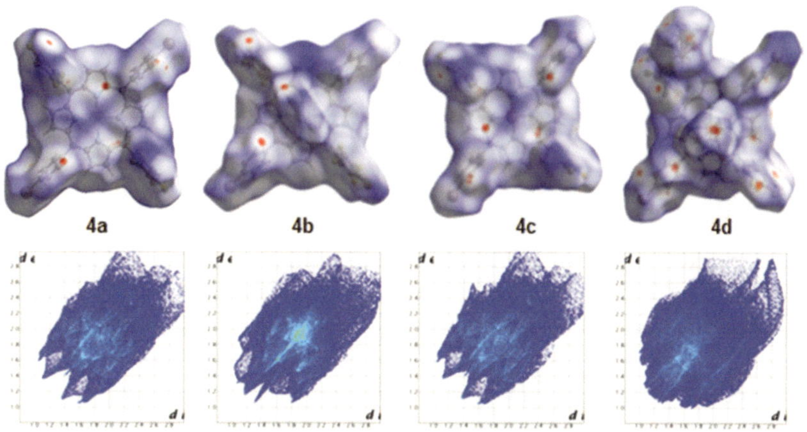

Figure 13. Hirshfeld surfaces with d_{norm} mapped ranging from -0.11 (blue) to 1.89 (red) and 2D FPs for the various intermolecular interactions in **4a–4d.**

The prominent spikes in the FP plots of **4a–4d** are associated with close contacts involving halogen atoms (F/Br) and can be attributed to F•••H/F/C, Br•••H/F interactions. It is clear from the analysis that the interactions involving halogens are the major contributors to the total HS area and the relative contributions are 62, 61, 66 and 57%, respectively, for **4a–4d**. Among these, 26–33% of close contacts are associated with F•••H interactions. Both symmetrical F•••F (~2.81 Å) as well as unsymmetrical F•••Br (~3.28 Å) halogen interactions are seen in all porphyrins except that the F•••Br interaction is absent in the case of **4b**. For porphyrins **4b** and **4d**, the F•••F/Br contact distances are broader in nature compared to **4a** and **4c** with no red spots on the HSs. Moreover, there is no F•••Br interaction in the case of iron(II) porphyrin **4b** and none of the porphyrins show Br•••Br interaction.

The C•••H contact is seen as peripheral wing like structure in all the four porphyrins and is associated with 13–18% of the total HSs. The H•••H interactions are more pronounced in iron and zinc derivative (21–22%), and it contributes only 7–10% for the isostructural pair **4a** and **4c**. The smallest fingerprint contributions occur for N•••H and C•••C. A detailed relative contribution of the different interactions to the HS was of porphyrins **4a–4d** is given in **Figure 14**.

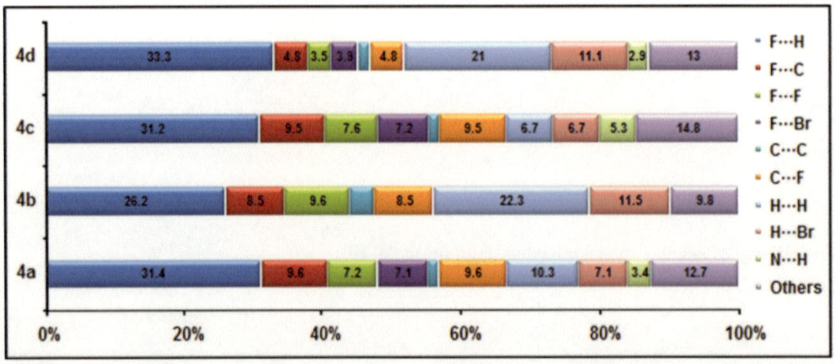

Figure 14. Percentage contribution of non-covalent interactions in porphyrins, **4a–4d** on the basis of HSs.

4. Conclusions

In conclusion, we have demonstrated the results of experimental crystallographic studies combined with computational HS analysis of a series of fluorinated porphyrins, **1–4**. The obtained results disclose that the isostructural couples exhibit similar kind of crystal packing in solid state with comparable intermolecular interactions. The HS analysis of compounds **1–4** indicates that the back bone of the net supramolecular arrangements are dictated by close contacts involving organic fluorine as well as C•••H and the cooperativity of weak H•••H interactions.

Acknowledgements

SS (SR/WOS-A/CS-146/2011) and CA (SR/FT/CS-25/2011, SB/EMEQ-016/2013 and FRG 10-11/0103) thank DST, New Delhi and NIT Calicut for research funding. The authors thank the students of Bioinorganic Materials Chemistry Laboratory, Department of Chemistry, National Institute of Technology Calicut, namely Rahul Soman, Fasalu Rahman Kooriyaden and Ramesh J for their contributions. We would like to thank Dr. Shibu M. Eappen, STIC, CUSAT, Kochi and Dr. Babu Varghese, SAIF, IIT Madras for the single crystal data collection and structure solution, refinement respectively. The authors also thank Prof. T. N. Guru Row and Mr. Vijith kumar, Solid State and Structural Chemistry Unit, Indian Institute of Science, Bangalore for their help to get one of the single crystal X-ray data collections.

Author details

Subramaniam Sujatha and Chellaiah Arunkumar*

*Address all correspondence to: arunkumarc@nitc.ac.in

Bioinorganic Materials Chemistry Laboratory, Department of Chemistry, National Institute of Technology Calicut, Kozhikode, Kerala, India

References

[1] Hoard JL: In Smith KM, editor. Stereochemistry of Porphyrins and metalloporphyrins. New York: Elsevier; 1975, pp. 317–380.

[2] Rothemund P: A new porphyrin synthesis. The synthesis of porphin. J. Am. Chem. Soc. 1936;58:625–627. doi:10.1021/ja01295a027

[3] Adler AD, Longo FR, Kampas F, Kim J: On the preparation of metalloporphyrins. J. Inorg. Nucl. Chem. 1970;32:2443–2445. doi:10.1016/0022-1902(70)80535-8

[4] Adler AD, Longo FR, Finarelli JD, Goldmacher J, Assour J, Korsakoff L: A simplified synthesis for *meso*-tetraphenylporphin. J. Org. Chem. 1967;32:476. doi:10.1021/jo01288a053

[5] Lindsey JS, Wagner RW: Investigation of the synthesis of *ortho*-substituted tetraphenylporphyrins. J. Org. Chem. 1989;54:828–836. doi:10.1021/jo00265a021

[6] Lindsey JS, Schreiman IC, Hsu HC, Kearney PC, Marguerettaz AM: Rothemund and Adler–Longo reactions revisited. Synthesis of tetraphenylporphyrins under equilibrium conditions. J. Org. Chem. 1987;52:827–836. doi:10.1021/jo00381a022

[7] Maldotti A, Amadelli R, Bartocci C, Carassiti V, Polo E, Varani G: Photochemistry of Iron-porphyrin complexes. Biomimetics and catalysis. Coord. Chem. Rev. 1993;125:143–154. doi:10.1016/0010-8545(93)85014-U

[8] Chou JH, Kosal ME, Nalwa HS, Rakow NA, Suslick KS: Applications of porphyrins and metalloporphyrins to materials chemistry. Urbana: Academic Press; 2000.

[9] Sternberg ED, Dolphin D, Brickner C: Porphyrin-based photosensitizers for use in photodynamic therapy. Tetrahedron. 1998;54:4151–4202. doi:10.1016/S0040-4020(98)00015-5

[10] Rashid H, Umar MN, Khan K, Anjum MN, Yaseen M: Synthesis and relaxivity measurement of porphyrin-based magnetic resonance imaging (MRI) contrast agents. J. Struct. Chem. 2014;55:910–915. doi:10.1134%2FS0022476614050163

[11] Ethirajan M, Chen Y, Joshia P, Pandey RK: The role of porphyrin chemistry in tumor imaging and photodynamic therapy. Chem. Soc. Rev. 2011;40:340–362. doi:10.1039/B915149B

[12] Ojima I, editor. Fluorine in medicinal chemistry and chemical biology. UK: Wiley-Blackwell; 2009.

[13] DiMagno SG, Biffinger JC, Sun H: Fluorinated porphyrins and corroles: Synthesis, electrochemistry, and applications. Fluor. Heterocycl. Chem. 2014;1:589–620. doi: 10.1007/978-3-319-04346-3_14

[14] Berger R, Resnati G, Metrangolo P, Weberd E, Hulliger J: Organic fluorine compounds: A great opportunity for enhanced materials properties. Chem. Soc. Rev. 2011;40:3496–3508. doi:10.1039/C0CS00221F

[15] Desiraju GR. Crystal engineering: The design of organic solids. Amsterdam: Elsevier; 1989.

[16] Chopra D, Guru Row TN. Role of organic fluorine in crystal engineering. CrystEngComm. 2011;13:2175–2186. doi:10.1039/C0CE00538J

[17] Goldberg I. Crystal engineering of porphyrin framework solids. Chem. Commun. 2005;1243–1254. doi:10.1039/B416425C

[18] Spackman MA, Jayatilaka D. Hirshfeld surface analysis. CrystEngComm. 2009;11:19–32. doi:10.1039/B818330A

[19] Wolff SK, Grimwood DJ, McKinnon JJ, Turner MJ, Jayatilaka D, Spackman MA. Crystal Explorer 3.1 (2013), University of Western Australia, Crawley, Western Australia, 2005–2013. http://hirshfeldsurface.net/CrystalExplorer.

[20] McKinnon JJ, Mitchell AS, Spackman MA: Hirshfeld surfaces: A new tool for visualising and exploring molecular crystals. Chem. Eur. J. 1998;4:2136–2141. doi:10.1002/(SICI)1521-3765(19981102)4:11<2136::AID-CHEM2136>3.0.CO;2-G

[21] Spackman MA, McKinnon JJ: Fingerprinting intermolecular interactions in molecular crystals. CrystEngComm. 2002;4:378–392. doi:10.1039/b203191b

[22] Littler BJ, Ciringh Y, Lindsey JS: Investigation of conditions giving minimal scrambling in the synthesis of *trans*-porphyrins from dipyrromethanes and aldehydes. J. Org. Chem. 1999;64:2864–2872. doi:10.1021/jo982452o

[23] Adler AD, Longo FR, Kampas F, Kim J: On the preparation of metalloporphyrins. J. Inorg. Nucl. Chem. 1970;32:2443–2445. doi:10.1016/0022-1902(70)80535-8

[24] Altomare AG, Cascarano G, Giacovazzo C, Gualardi A: Completion and refinement of crystal structures with SIR92. J. Appl. Crystallogr. 1993;26:343–350. doi:10.1107/S0021889892010331

[25] Sheldrick GM. SHELXL97. Goettingen, Germany: University of Goettingen; 1997.

[26] Kadish KM, Araullo-McAdams C, Han BC, Franzen MM: Syntheses and spectroscopic characterization of (T(p-Me2N)F4PP)H2 and (T(p-Me2N)F4PP)M. J. Am. Chem. Soc. 1990;112:8364–8368. doi:10.1021/ja00179a021

[27] Schauer CK, Anderson OP, Eaton SS, Eaton GR: Crystal and molecular structure of a six-coordinate zinc porphyrin: Bis(tetrahydrofuran)(5,10,15,20-tetraphenylporphinato)zinc(II). Inorg. Chem. 1985;24:4082–4086. doi:10.1021/ic00218a024

[28] Tarahhomi A, Pourayoubi M, Golen JA, Zargaran P, Elahi B, Rheingold AL, Leyva Ramırezc MA, Percino TM: Hirshfeld surface analysis of new phosphoramidates. Acta Cryst. 2013;B69:260–270. doi:10.1107/S2052519213009445

[29] Bondi A: Van der Waals volumes and radii. J. Phys. Chem. 1964;68:441–451. doi:10.1021/j100785a001

[30] Batsanov SS: Van der Waals radii of elements. Inorg. Mater. 2001;37:871–885. doi: 10.1023/A:1011625728803

[31] Grabowsky S, Dean PM, Skelton BW, Sobolev AN, Spackman MA, White AH: Crystal packing in the 2-R,4-oxo-[1,3-a/b]-naphthodioxanes–Hirshfeld surface analysis and melting point correlation. CrystEngComm. 2012;14:1083–1093. doi:10.1039/c2ce06393j

[32] Soman R, Sujatha S, Arunkumar C: Quantitative crystal structure analysis of fluorinated porphyrins. J. Fluor. Chem. 2014;163:16–22. doi:10.1016/j.jfluchem.2014.04.002

[33] Reed CA, Mashiko T, Scheidt WR, Spartalian K, Lang G: High spin iron (II) in the porphyrin plane. Structural characterization of (mesotetraphenylporphinato) bis (tetrahydrofuran)iron (II). J. Am. Chem. Soc. 1980;102:2302–2306. doi:10.1021/ja00527a028

[34] Schauer CK, Anderson OP, Eaton SS, Eaton GR: Crystal and molecular structure of a six-coordinate zinc porphyrin: Bis(tetrahydrofuran) (5,10,15,20- tetraphenylporphinato) zinc(II). Inorg. Chem. 1985;24:4082–4086. doi:10.1021/ic00218a024

[35] Bhyrappa P, Arunkumar C: Structural and electrochemical properties of β-tetrabromo-mesotetrakis(4-alkyloxyphenyl)porphyrins and their metal complexes. J. Chem. Sci. 2010;122:233–238. doi:10.1007/s12039-010-0027-6

[36] Bhyrappa P, Arunkumar C, Varghese B, Sankara Rao DS, Prasad SK: Synthesis and mesogenic properties of β-tetrabrominated tetraalkyloxyporphyrins. J. Porphyrins Phthalocyanines. 2008;12:54–64. doi:10.1142/S108842460800008X

[37] Scheidt WR: Systematics of the stereochemistry of porphyrins and metalloporphyrins, Kadish KM, Smith KM, Guilard R, New York: Academic Press; 2000, vol. 3, pp. 49–112.

[38] Kooriyaden FR, Sujatha S, Arunkumar C: Synthesis, spectral, structural and antimicrobial studies of fluorinated Porphyrins. Polyhedron. 2015;97:66–74. doi:10.1016/j.poly.2015.05.018

[39] Soman R, Sujatha S, De S, Rojisha VC, Parameswaran P, Varghese B, Arunkumar C: Intermolecular Interactions in fluorinated tetraarylporphyrins: An experimental and theoretical study. Eur. J. Inorg. Chem. 2014;2653–2662. doi:10.1002/ejic.201402008

Section 2

Non-crystalline Solids

Chapter 5

Ultra-Thin Plasma-Polymerized Functional Coatings for Biosensing: Polyacrylic Acid, Polystyrene and Their Co-Polymer

Paola Rivolo, Micaela Castellino,
Francesca Frascella and Serena Ricciardi

Additional information is available at the end of the chapter

http://dx.doi.org/10.5772/62899

Abstract

Recently, many efforts have been done to chemically functionalize sensors surface to achieve selectivity towards diagnostics targets, such as DNA, RNA fragments and protein tumoural biomarkers, through the surface immobilization of the related specific receptor. Especially, some kind of sensors such as microcantilevers (gravimetric sensors) and one-dimensional photonics crystals (optical sensors) able to couple Bloch surface waves are very sensitive. Thus, any kind of surface modifications devoted to functionalize them has to be finely controlled in terms of mass and optical characteristics, such as refractive index, to minimize the perturbation, on the transduced signal, that can affect the response sensitivity towards the detected target species.

In this work, the study and optimization of ultra-thin plasma polymers and copolymers, compatible with these constrains and obtained from the vapours of acrylic acid containing a carboxylic (–COOH) group and styrene (an aromatic molecule with a vinyl as substituent at the ring), are reported. The obtained plasma polyacrylic acid (PPAA), plasma polystyrene (PPST) and their copolymer (PPAA–ST), characterized through optical contact angle analysis (OCA), Fourier transform infrared (FTIR) spectroscopy in attenuated total reflection (ATR-FTIR), X-ray photoelectrons spectroscopy (XPS), and atomic force microscopy (AFM), are shown to match specific and critical requirements, such as low thickness (~40 nm) and refractive index (~1.5), high surface density of reactive groups (10^{15}–10^{16} COOH/cm^2), bio-antifouling properties where required, reproducibility, and chemical resistance and stability.

Keywords: plasma polymerization, acrylic acid, styrene, thin films, biosensing

1. Introduction

In the last decades, one of the most critical issues in biosensing devices was the proper surface chemical modification aimed to covalently immobilize a bioreceptor (e.g. a capture antibody) or probe (a single-stranded DNA, ss-DNA), able to selectively bind a target, by matching an antigen or an oligonucleotide sequence (DNA or RNA fragment) of interest for biodiagnostics, respectively [1–3]. Polymeric thin coatings, such as polyacrylic acid, polydi(ethylene glycol) monovinyl ether, maleic anhydride, obtained by a plasma-enhanced chemical vapour deposition (PECVD) [4–8], at low-pressure conditions, are quite suitable for the chemical functionalization of a sensing surface [9, 10], owing to the following reasons: (i) they do not change the chemical properties of the bulky material of a sensing device so leaving unaffected its working principle; (ii) they are conformal and (iii) do not need particular chemistry exposed to the sensor surface for adhesion; and (iv) they can show different features in terms of density of functional groups, wettability (hydrophilicity/hydrophobicity) and chemical stability by simply varying the process parameters (modulation of the plasma discharge, monomer or mixture of monomers/process gas, reactant vapour partial pressure, etc.), even though the same reagent is used as precursor (acrylic acid, vinyl acetic acid, allyl alcohol, allylamine) [11–13].

The chemical stability, in aqueous solution or other solvents (according to the biological protocol/technological procedure), is one of the mandatory goal to achieve, dealing with the plasma polymers exposing monotype functional groups suitable for covalent binding, as reported for polyacrylic acid and polyallylamine [14–19]. To enhance this last aspect, plasma copolymerization can be adopted to better tune the distribution and density of functional groups by mixing the monomer containing the functional groups (e.g. acrylic acid, allylamine, allyl alcohol) of interest with a precursor which does not contain specific functionalities (e.g. ethylene, styrene) and which has the function to dilute the reactive groups at the surface [20]. This last kind of monomers can be used alone to get a plasma polymer with bio-antifouling properties so allowing, for reference purpose, the detection of a blank response entirely unaffected by unspecific adhesion of biological materials, as shown by Sardella et al. for surface micro-domains of plasma-polymerized polyacrylic acid and polyethylene oxide [21]. The described polymers are also compatible with microscale lateral patterning based on photolithography combined with multiple subsequent plasma deposition steps. [22]

Concerning high-performing sensors, such as gravimetric (microcantilevers) [23] or optical (Bloch surface waves (BSW)-based one-dimensional photonic crystals) [24–27] microdevices, the limited thickness of the functionalizing coating represents a crucial task to reach in order to avoid losses of sensitivity in the biodetection, as, for the first type, additional masses have to be minimized and, for the second type, the perturbation of refractive index at the surface of the sensing device has to be finely tuned.

In this work, the study, optimization and characterization will be presented of ultra-thin plasma polymers and copolymers obtained from the vapours of acrylic acid containing a carboxylic (–COOH) group and styrene (an aromatic molecule with a vinyl as substituent at the ring). A particular attention will be devoted to match specific and critical requirements, such as low thickness and refractive index, high surface density of reactive groups, bio-antifouling properties where required, reproducibility, and chemical resistance and stability.

Several plasma process conditions, amount ratio of the two kinds of precursors and thicknesses, ranging from 10 to 50 nm, were considered and the related coatings were characterized by means of OCA, ATR-FTIR, XPS and AFM. Through these techniques, it was possible to assess the chemical composition of the obtained films, the hydrophilicity and hydrophobicity, the effect of solvents on the chemical structure and thickness of coatings due to eventual dissolution.

The −COOH surface density of plasma-polymerized acrylic acid (PPAA) and copolymer (PPAA–ST) obtained by mixing styrene moieties to acrylic acid vapours during the plasma process was quantified by means of colorimetric titration through toluidine blue O (TBO) [28, 29].

Finally, the surface activity, towards test fluorophore-labelled proteins, versus binding or antifouling properties of the three selected polymers (PPAA, PPST and their copolymer) was studied by fluorescence microscopy.

2. Experimental

2.1. Chemicals

Pure acrylic acid, anhydrous, 99% (vapour pressure = 3.1 Torr at 20°C and 1 atm) and pure styrene, ≥99% (vapour pressure = 5 Torr at 20°C and 1 atm) liquid monomers, acetone, ACS reagent, ≥99.5% and acetic acid (glacial, ≥99%) were purchased by Sigma Aldrich. TBO (technical grade) and sodium hydroxide were obtained from Fluka. The AlexaFluor546-conjugated protein A (PtA-AF546) was purchased from Invitrogen. The phosphate buffer saline (PBS) 1× solution was from GIBCO®. All water solutions were prepared with Milli-Q™ grade water (Merk-Millipore, Milan).

2.2. Plasma polymerization system

Plasma treatments were carried out at room temperature in a low-pressure plasma-enhanced CVD reactor (chamber base pressure = 28 mTorr; RF = 13.56 MHz) purchased from IONVAC s.r.l., Rome, Italy [30]. The system is equipped with a delivery frame suitable to inject organic vapours coming from liquid reactants (monomeric precursors) stored in reservoirs intercepting the carrier gas lines.

The parameters of the plasma processes will be reported and discussed along the text in the Findings paragraph.

2.3. Characterization techniques

OCA measurements were used to investigate the surface properties of the differently functionalized samples in terms of hydrophilicity and hydrophobicity The OCAH 200 instrument

(DataPhysics Instruments GmbH), used for the characterization, is equipped with a charge-coupled device (CCD) camera and an automatic dosing system for the liquids. The selected ones for the sessile droplet method in static mode analysis (droplet volume = 1.5 μl) were deionized water MilliQ grade (H_2O) and diiodomethane (CH_2I_2, Sigma Aldrich). The Young–Laplace method was applied to fit the drop profiles and the SCA20 software was used to calculate contact angles between fitted function and base line. A data set of at least three droplets for each liquid was used to evaluate the standard deviation for each kind of sample.

ATR-FTIR spectra were collected in the range 4000–600 cm^{-1}, at 4 cm^{-1} resolution. Sixty-four scans were accumulated for each spectrum on a Nicolet 5700 FTIR spectrometer (ThermoFisher) equipped with a room temperature working deuterated triglycine sulphate (DTGS) detector and a diamond crystal ATR accessory.

A PHI 5000 Versaprobe scanning X-ray photoelectron spectrometer (monochromatic Al Kalpha X-ray source with 1486.6-eV energy) was employed to check the material surface chemistry. High resolution (pass energy: 23.5 eV) and survey spectra (pass energy: 187.85 eV) have been collected using a beam size of 100 μm. A combination of an electron gun and an Ar ion gun has been used as a neutralizer system to compensate the positive charging effect during the analysis, due to not perfectly conductive surfaces. Fitting procedure and deconvolution analysis have been done using the Multipak 9.6 dedicated software. All core-level peak energies were referenced to C1s peak at 284.5 eV (C–C/C–H sp^2 bonds).

AFM measurements were performed by the WiTec AlphaSNOM system that implements a combined AFM and scanning near-field optical microscopy (SNOM) in collection/illumination and scattering mode by employing microfabricated, hollow pyramidal probes. The profile step measurements by AFM module of the deposited plasma polymers were collected by tapping mode.

Wide-field fluorescence microscopy was performed by a LEICA DM-LM microscope (objective 20×) equipped with a Hg vapour arc lamp (50 W) and L5-type combination filters suitable for fluorescein-like fluorophore. The coated samples (Si and SiO_2) were incubated with a 0.1 mg/ml proteinA, Alexa Fluor 546 aqueous solution (dispensed volume = 10 μl) and rinsed in PBS, pH 7.3 + H_2O to remove unbounded protein.

Colorimetric titration of –COOH functionalities of PPAA and PPAA–ST films was performed by means of TBO. The amino group contained in TBO molecule reacts specifically with a surface carboxylic group according to a 1:1 ratio [28]. Functionalized samples (c-Si substrates were used) were contacted with 3 ml of 0.5 mM TBO aqueous solution (pH = 10) at 37°C, for 5 h, in the dark. Then samples were copiously rinsed with 0.1 mM NaOH solution to remove the unreacted dye. Afterwards, samples were transferred in 1.5 ml of 50% v/v acetic acid solution and shaken for 10 min, in order to completely release the TBO linked to the carboxylic functional groups. Colorimetric dosing of TBO (and consequently of carboxylic groups) was performed by using a 2100-C microplate reader (Ivymen Optic System), by recording the optical densities (OD) of the solutions recovered after the reaction with the samples at 630 nm. A solution of 50% v/v acetic acid was used as reference background. By applying the Lambert–Beer law to the OD recorded for the different samples, on the basis of the comparison with an internal calibration obtained with TBO solutions (at known concentrations and volumes), and through the normalization for the geometric surface of the test samples, the evaluation of carboxylic functionalities density (in terms of groups/cm^2) can be performed.

In order to get information on the chemical characteristics of PPAA, PPST and PPAA–ST films, flat substrates, that is silicon (CZ/1-0-0, boron/P-type), Corning (Danville) glass and polyethylene (low-density PE; thickness 2.3 mm; Good Fellow) were used for the most of characterizations. For OCA, the different substrates were used in order to verify that the coating has the same properties without any dependence on substrate chemical composition, even for low thicknesses. For FTIR-ATR spectroscopy, silicon was not used owing to its refractive index higher than the ATR module diamond crystal. PE and Corning glass substrates have absorptions in different spectral range hiding thin film vibrational absorption features in different complementary parts of the selected range of spectra collection. The use of both substrates allows the investigation of the plasma polymer vibrational modes in the whole 400–4000 cm^{-t}. For XPS and AFM, only silicon was used.

3. Findings

3.1. Plasma polymerized acrylic acid

The advantages of using acrylic acid as a precursor for plasma polymerization are the quite low vapour pressure that makes controllable the flux of vapours by manual metering valve (calculated AA vapour flow was approximately of 3 sccm), the nearly absent toxicity of reactants, the presence of an unsaturation (vinylic double bond) allowing the radical polymerization mechanism and a carboxylic functionality (–COOH).

This chemical group is particularly suitable to biosensing applications because it forms easily amidic bridges by reacting with amino groups of biomolecules (e.g. receptor antibodies) with (binding efficiency is enhanced) or without further chemical activation.

In previous works of the authors [4, 26, 30, 31], films with thicknesses included between 80 and 120 nm (processes performed by T_{Dep} between 10 and 20 min) and prepared by means of both continuous wave (CW) and modulated wave (MW) plasma discharge were widely studied. The coating obtained by CW discharge were shown to not retain enough carboxylic functionalities while, among MW coatings, only the film obtained with an RF starting power of 200 W, a duty cycle of 0.1 with a t_{on} of 10 ms and a t_{off} of 90 ms with Ar as carrier gas, were optimal in terms of stability and reproducibility. A step of immersion in d-H_2O is required after plasma synthesis in order to remove surface oligomers (left at the end of plasma process) so making the film chemically stable for the following interaction steps (biosensing) in aqueous media. Moreover, to enhance the adhesion among the substrate and the deposited film an Ar plasma treatment was required before the film deposition. Thick films allowed a complete and reliable surface characterization that consisted of OCA, FTIR-ATR spectroscopy, XPS analysis, fluorescence microscopy, quartz microbalance system and colorimetric titration. By means of the last, an exceptional value of 1.15 ± 0.35 × 10^{16}groups/cm^2 was found. Moreover, from ellipsometric characterization, it also aroused that the optimal film has a refractive index of 1.51 measured on 40-nm thick layers, after d-H_2O rinsing and from electrokinetic characterization based on the zeta (ζ) potential measurements, that allows the evaluation of acid–base properties, came that PPAA isoelectric point is at pH = 2.1. This indicates that the surface charge originates from the dissociation of the carboxylic acid groups of the PPAA chains and so from the deprotonation of –COOH species [4].

In the following, the characterization of PPAA ultra-thin films is reported in order to define the minimum thickness that can be reached without affecting the chemical resistance mainly in aqueous media and the functional group number and reactivity.

For these processes, Argon (Ar) is the AA vapour–carrier gas (flow =20 sccm). It is made bubble in liquid AA in order to enhance vapours formation and transport them into the chamber. The in-chamber total pressure in presence of the Ar–AA mixture is about at 220–240 mTorr, with a partial pressure of AA vapours of about 25 mTorr. The plasma parameters were an RF starting power of 200 W, a duty cycle of 0.1 with a t_{on} of 10 ms and a t_{off} of 90 ms.

After the film deposition, characterizations were performed before and after rinsing the substrates in de-ionized water (d-H_2O) for 20 min.

OCA angles performed on bare silicon (Si) and Corning glass with d-H_2O and CH_2I_2 (diiodomethane) and related calculation of surface energy (SE, W_{sl}) divided in its disperse (W^d_{sl})and polar (W^h_{sl}) components, are reported in **Table 1**.

Substrate	OCA_{H_2O} (deg)	$OCA_{CH_2I_2}$ (deg)	W_{sl} (mN/m)	W^d_{sl} (Watt)	W^h_{sl} (min)
Si	47 ± 1	46 ± 3	53,24	25,29	27,95
Corning	54 ± 1.5	38 ± 2	50,88	30,94	19,94

Table 1. OCA angles and SE data for bare substrates.

Time (min)	OCA_{H_2O} (deg)	$OCA_{CH_2I_2}$ (deg)	W_{sl}(mN/m)	W^d_{sl} (mN/m)	W^h_{sl} (mN/m)
2.5	15±1	34±2	70,89	27.17	43.72
4	11.3±0.9	35±1	71.80	26.56	45.24
5	17.1±0.8	32.3±0.5	70.41	28.24	42.17
7.5	15.6±0.2	31.2±0.4	70.98	28.63	42.35
10	15±1	34±2	70.94	27.20	43.74

Table 2. OCA angles and SE data for different PPAA films deposited at increasing T_{Dep} on c-Si.

Time (min)	OCA_{H_2O} (deg)	$OCA_{CH_2I_2}$ (deg)	W_{sl}(mN/m)	W^d_{sl} (mN/m)	W^h_{sl} (mN/m)
2.5	16.7 ± 0.4	36 ± 0	70.27	26.63	43.64
4	14.2 ± 0.6	35.0 ± 0.9	71.09	26.89	44.20

http://dx.doi.org/10.5772/62899

Time (min)	OCA_{H_2O} (deg)	$OCA_{CH_2I_2}$ (deg)	W_{sl}(mN/m)	W^d_{sl} (mN/m)	W^h_{sl} (mN/m)
5	17.4 ± 0.9	35 ± 2	70.09	26.99	43.10
7.5	17.2 ± 0.1	34.9 ± 0.4	70.18	27.14	43.04
10	16.6 ± 0.6	36 ± 2	70.26	26.33	43.92

Table 3. OCA angles and SE data for different PPAA films deposited at increasing T_{Dep} on Corning glass.

An example of the similar surface chemistry of PPAA at the lower T_{Dep} for both c-Si and Corning glass is reported in **Figure 1**. It also shows that any effect of substrate chemistry is absent. The high polar component of the surface energy (see **Tables 2** and **3**) confirms that the surface of the test substrates is homogeneously coated by polar hydrophilic species.

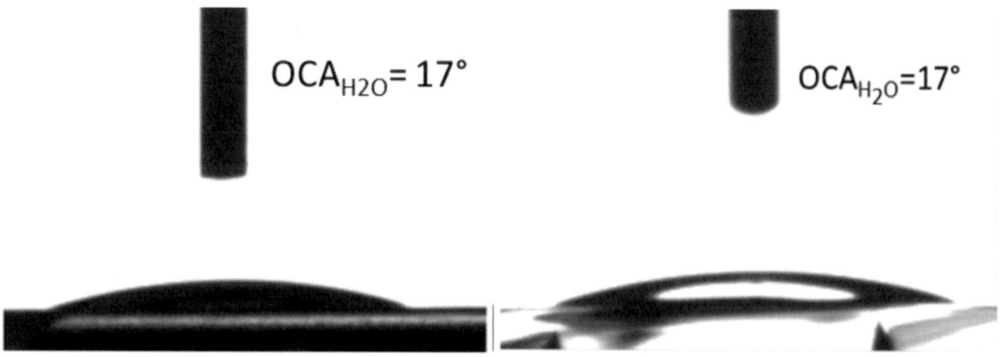

Figure 1. Example of OCA_{H_2O} on PPAA-2.5 min film on c-Si and on Corning.

The FTIR-ATR spectra of **Figure 2**, collected on both PPAA-deposited PE and Corning glass, confirm that the species exposed at surface are carboxylic groups and that they are also visible for the thinner film.

An effect of thickness growth is evident. The absorption features of –COOH groups increase according to the length of T_{Dep}: the band between 3750 and 3200 cm^{-1} assigned to the absorption envelope of O–H stretching mode, the C–H stretching mode at 2980–2870 cm^{-1} visible on Corning glass substrate, at 1730–1710 cm^{-1} the absorption related to C=O stretching, corresponding to both carboxylic and carbonyl functionality, the C–H bending mode at 1460–1370 cm^{-1} and the C–O stretching mode at 1200 cm^{-1} visible on PE substrates. The effects of rinsing in water are also visible (dotted curves) as a slight decrease of –COOH absorption features for higher thicknesses.

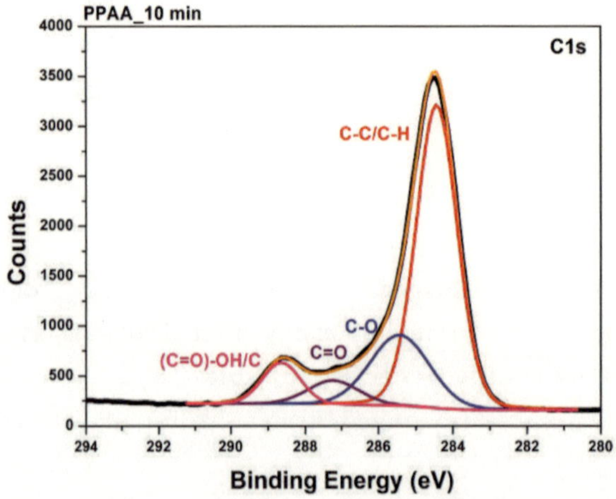

Figure 2. FTIR-ATR spectra measured on PE (a) and Corning glass (b) substrates coated with PPAA films obtained at increasing T_{Dep}.

XPS analysis (HR–C1s spectrum, reported in **Figure 3**) of the PPAA coating, deposited on c-Si, shows the C–C/C–H peak at 284.5 eV, a second intense component at 285.4 eV, assigned to C–O species, a third one at 287.3 eV related to C=O groups and a third component at 288.7 eV related to the (C=O)O–H/C species. All the features are consistent with the presence of carboxylic chemical groups and they will be used for the sake of comparison in the paragraph devoted to XPS analysis on PPAA–ST films.

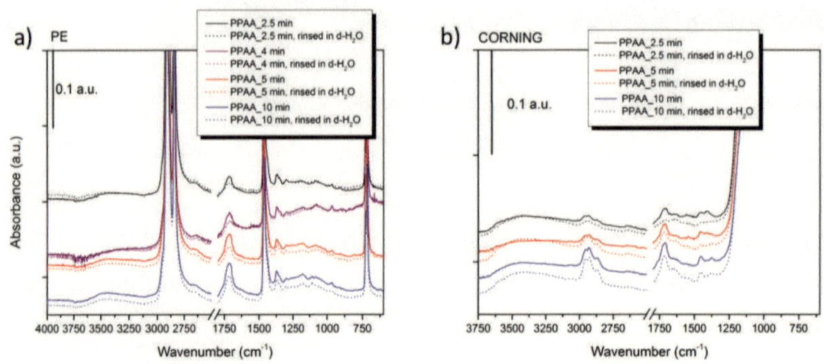

Figure 3. XPS analysis on PPAA obtained TDep of 10 min (C1s HR spectrum).

In the following, the correlation between T_{Dep} and film thickness in the range 0–50 nm was performed by means of AFM for profilometric purposes, in order to get a calibration of selected process and information about deposition rate.

From **Table 4** arises that the deposition rate has not a linear behaviour. For low times of deposition, the rate is not reproducible (several processes at the same deposition time were performed) due the starting step of pulsed discharge stabilization that is not controllable.

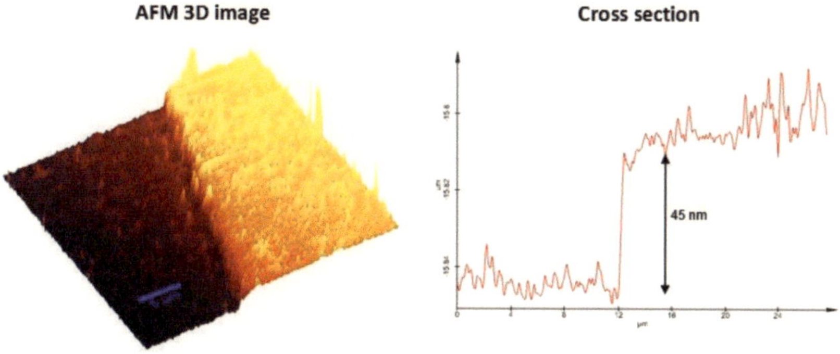

Figure 4. PPAA (T_{Dep} = 5 min) AFM characterization.

For this reason, the minimum deposition time allowing reproducible thicknesses is 4 min, that is a thickness of 40 nm.

Time dep (min)	Time dep (s)	Thickness (nm)	Dep rate (nm/s)	Dep rate (nm/min)
2.5	150.0	25.0	0.17	10
4	240.0	40.0	0.17	10
5 (Figure 4)	300.0	45.0	0.15	9
7.5	450.0	60.0	0.13	8
10	600.0	85.0	0.14	8.5

Table 4. AFM profilometry: PPAA thickness values at different deposition times.

3.2. Plasma-polymerized styrene

Styrene is an aromatic molecule (C_6H_6–CH=CH_4) which, by decomposing, during plasma process, and reassembling in form of cross-linked chains, allows to get a polymer with mechanical toughness, thermal stability, dielectric properties and chemical resistance/ inertness. These properties make it a good bio-antifouling surface. Moreover, it can be used to dilute other functional monomers, thus allowing to introduce aliphatic fragments/aromatic rings among carboxylic groups arising from a chemically functional monomer, such as acrylic acid. A plasma copolymer can be so obtained, as reported in the next paragraph.

The explored processes, by changing different parameters such as the discharge mode (CW and MW), were performed at 2 and 10 min of deposition time. With respect to AA, the monomer flow, measured when the manual valve intercepting the reservoir is open, is not easily controllable and it is highly sensitive to external temperature variations. This could create problems for the reproducibility of the processes and does not allow the measure of the vapour flow of styrene as in the case of acrylic acid.

Process type	P_{RF} (Watt)	D.C.	t_{on} (ms)	t_{off} (ms)	P_{AVE} (Watt)
PPST_1	60	CW			
PPST_2	200	0.1	20	180	20
PPST_3	200	0.1	50	450	20
PPST_4	200	0.1	10	90	20
PPST_5	200	0.5	50	50	100

Table 5. PPST process parameters.

XPS analysis was carried out on selected spectra, obtained with time of deposition of 10 min, in order to confirm the presence of intact benzene ring and to evaluate the content of oxygen that, even if unwanted, seems to be embedded in the coatings (due to quite low vacuum produced by the chamber pumping system). This side effect is supported by data reported in literature.

The analysis has been performed on the MW-obtained coatings only. **Figure 5** reports the C1s HR spectrum (a) and the O1s HR spectrum (b) of process PPST_2 of **Table 5**, obtained at P = 200W; DC = 0.1; t_{on}/t_{off} = 20 ms/180 ms. Peaks have been deconvoluted using Gaussian–Lorentz functions, while backgrounds have been removed with Tougaard functions. Fitting peaks are:

- the lower energy peak (284.3 eV) has been assigned to ring carbons;,
- the peak at 284.5 eV has been attributed to the presence of C atoms linked as C–C/C–H,
- the peak at 285.5 eV is commonly attributed to C–O–C,
- the higher energy peak (~291 eV) is due to π–π^* transitions arising from the presence of the aromatic rings. It is called "shake-up" satellite because related to "shake-up" excitations taking place in the π orbitals on the benzene rings.

Concerning the O1s HR spectrum, only one fitting peak was found, related to C–O–C chemical bonds.

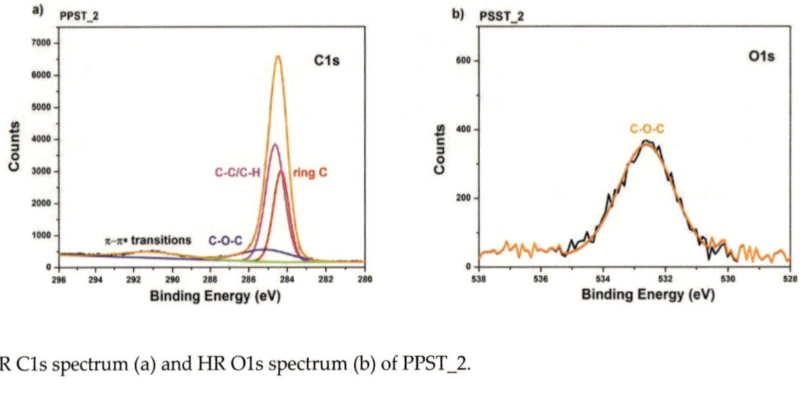

Figure 5. HR C1s spectrum (a) and HR O1s spectrum (b) of PPST_2.

http://dx.doi.org/10.5772/62899

The processes of the PPST_2 kind are the most reproducible, without a dependence on the deposition time, for this reason it is the most representative to single out the features of the HR C1s and O1s spectra of all the PPST processes.

For all processes, XPS peaks, in the energy range corresponding to valence band photoelectrons, are similar to that of a standard polystyrene. In fact, an extensive amount of information is also available from the valence band region. This is because the shifts in this region arise from changes in chemical bonding, rather than the chemical shifts caused by the environment found in the core region.

Process type	Dep time (min)	O/C atomic percentage ratio
PPST_1	10	0.07
PPST_2	2	0.05
PPST_2	10	0.04
PPST_3	10	0.04
PPST_4	10	0.02
PPST_5	10	0.04

Table 6. O/C atomic ratio referred to elements percentage obtained from survey spectra of PPST films.

From **Table 6** comes that for all processes the content of oxygen is quite negligible. **Figure 6a** shows that, for the most stable process in terms of water resistance, the energy band affected by the rinsing in water is the one relate to ring carbons, so suggesting that the chemical weakness of PPST is the cross-linking among the benzene rings. However, this effect does not affect the hydrophobic/antifouling properties of the PPST coating.

Figure 6b reports the intensity distribution of components of C1s spectrum of three selected processes (**Table 6**). The comparison puts in evidence that the process at $t_{on}/t_{off} = 20/180$ produces a higher amount of ring carbons with respect to the others.

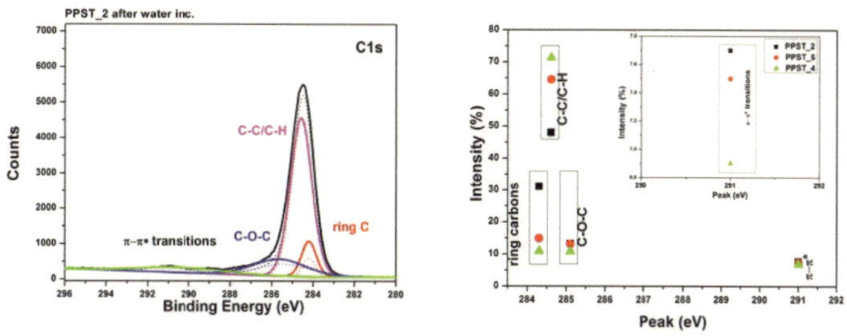

Figure 6. (a) Effects of water rinsing on component fitting peaks of HR-C1s spectrum of coating PPST_2 (dot line for water incubated sample, straight line for not-treated sample). (b) Intensity distribution of HR C1s components for sample PPST_2, PPST_4 and PPST_5 samples.

The process PPST_2 10 min; P = 200W; DC = 0.1; 20 ms/180 ms is the most performing in terms of low grade of oxidation, good retention of aromatic chemistry that ensures inertness and discourage electrostatic interaction.

To investigate on the bio-antifouling properties of it, it was deposited on the silica (SiO_2) surface of a glass substrate (a sector of the sample was masked) and incubated for 30 min with a solution of PtA labelled with a fluorescent marker (Alexa Fluor 546). After rinsing and drying under N_2 flux, according to the "Experimental" Section, the sample was observed by fluorescence microscopy.

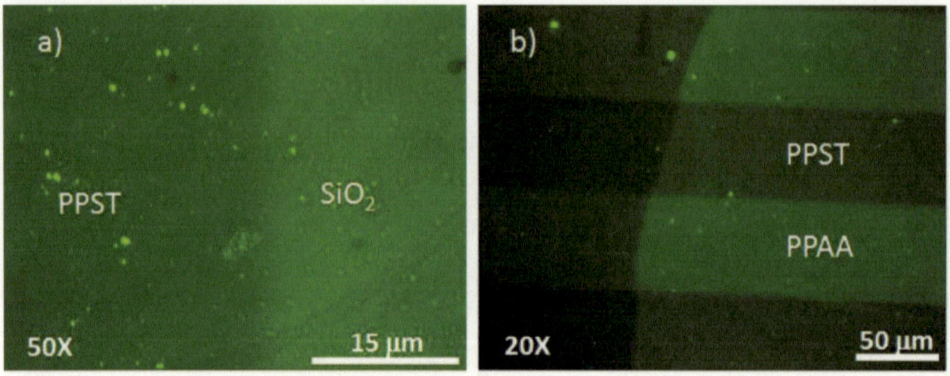

Figure 7. Fluorescence micrographs of PtA-Alexa Fluor 546 adhesion: (a) comparison between PPST film and bare SiO_2, (b) comparison between PPST and PPAA films.

It is evident from **Figure 7a** that selected PPST coating strongly limits unspecific adsorption of biomolecules, as fluorescent emission from PPST is lower or nearly absent with respect to the one coming from SiO_2. Moreover, concerning the lateral patterning involving PPST and PPAA is evident from **Figure 7b** that PPAA is highly PtA chemisorbing and PPST is bio-anti-fouling.

Also for PPST such as for PPAA is mandatory to verify that the compositional characteristics are retained also for thin layer. For this reason, the OCA measurements were performed on coatings obtained by different deposition time on the standard test substrates: Si and Corning glass.

The measured angles and related surface energies for coated Si and Corning substrates are reported in **Tables 7** and **8**, respectively. OCA_{H_2O} are quite high and confirm the hydrophobic properties of all the films. But a slight decrease of it is observed by incrementing the deposition time (here expressed in seconds). This trend suggests that longer is the deposition time, larger is the side oxidation effect produced by the process, and that is confirmed by FTIR-ATR spectra. Moreover, it is confirmed by the increase of the polar component of surface energy W^h_{sl}.

Time (s)	OCA_{H_2O} (deg)	$OCA_{CH_2I_2}$ (deg)	W_{sl} (mN/m)	W^d_{sl} (mN/m)	W^h_{sl} (mN/m)
15	77.2 ± 1.9	9.6 ± 0.8	50.15	47.11	3.04
120	77.3 ± 1	14.6 ± 0.8	49.27	46.1	3.17
240	76.5 ± 0.8	19.5 ± 1	48.15	44.43	3.72
600	72.8 ± 1.7	20.4 ± 0.9	48.52	42.86	5.66

Table 7. OCA and SE measurements performed on Si substrates coated with PPST films according to process PPST_2, at different deposition times.

Time (s)	OCA_{H_2O} (deg)	$OCA_{CH_2I_2}$ (deg)	W_{sl} (mN/m)	W^d_{sl} (mN/m)	W^h_{sl} (mN/m)
15	78.3 ± 0.9	21.4 ± 1.3	47.47	44.29	3.18
120	77.0 ± 1	14.2 ± 1.5	49.37	46.1	3.27
240	76.1 ± 1.3	16.3 ± 1.3	48.97	45.29	3.68
600	70.3 ± 1.7	17.1 ± 1.4	49.69	43.38	6.31

Table 8. OCA and SE measurements performed on Corning glass substrates coated with PPST films according to process PPST_2 at different deposition times.

OCA measurements were performed also on films after acetone rinsing, in view of photolithographic approaches for lateral patterning which involve the production of alternated chemical domains PPAA/PPST. OCA_{H_2O} slightly increases (see **Figure 8**) but the films chemically resists to the treatment.

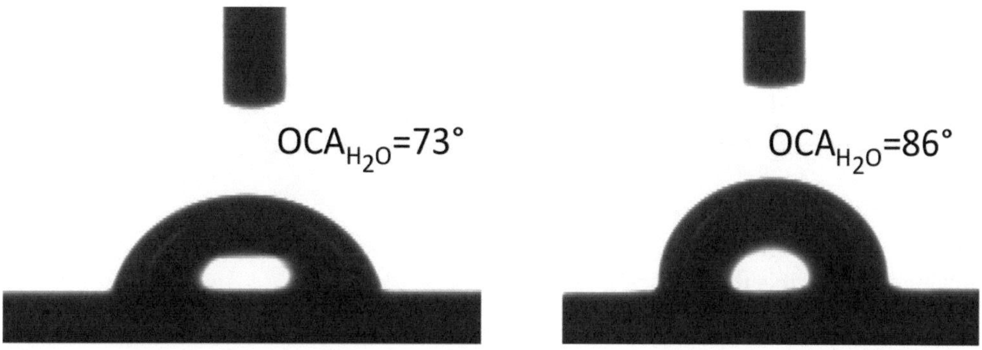

Figure 8. Comparison of OCA_{H_2O} on Si-PPST_2; P = 200 W; DC = 0.1; 20 ms/180 ms –600 s (10 min) process before (left) and after (right) rinsing in acetone.

The FTIR-ATR spectra of **Figure 9**, collected on the OCA characterized samples, on both PPST deposited PE and Corning, show the band associated to the C=C stretching in the aromatic ring at 1600 cm^{-1} and below 1200 cm^{-1}, the features due to the aliphatic backbone chain absorption modes. Most of spectra are present between 1730 and 1645 cm^{-1} absorption due to slight oxidation occurring during the process, and only for the thicker films, at 2980–2870 cm

$^{-1}$, the signals related to aliphatic C–H stretching mode are visible. An effect of thickness growth is observed as a slight increase of all described fingerprints. Finally, an effect of enhancement of aliphatic C–H stretching absorption, due to acetone rinsing is observed on the PPST_2-600s (10 min), coating (pink dashed curve), in accordance with OCA results (increase of hydrophobicity). Water rinsing seems to leave unaffected the PSST coating (related spectra are the dotted curves).

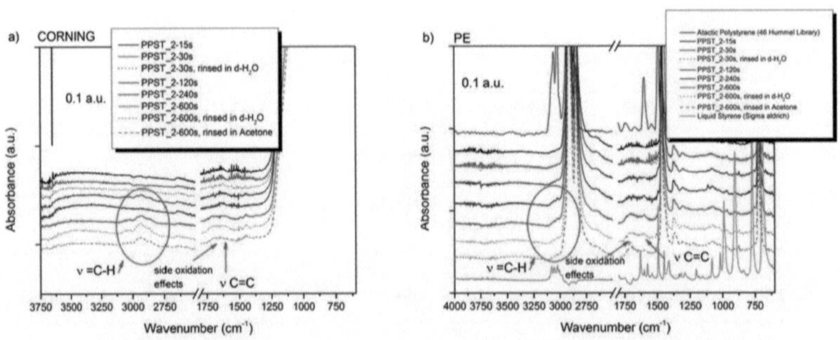

Figure 9. FTIR-ATR spectra measured on Corning (a) PE (b) substrates coated with PPST films obtained at increasing deposition time and compared with spectra of atactic standard polystyrene and liquid styrene.

In order to get a calibration of thicknesses of PPST in the lower range possible, according to deposition time, AFM was used with a profilometric approach and measured thicknesses are reported in **Table 9**.

Time dep (s)	Thickness (nm)	Dep rate (nm/s)	Dep rate (nm/min)
120	25.0	0.21	12.6
420	113.0	0.27	16.2

Table 9. AFM profilometry: PPST thickness values at different deposition times.

For T_{Dep} lower than 2 min, the rate is not reproducible due to the starting step of pulsed discharge stabilization that is not controllable. For this reason, the minimum deposition time allowing reproducible thicknesses is 120 s (2 min).

3.3. Copolymer

Plasma copolymerization is based on the activation of a plasma process in a gaseous mixture where the vapours of a monomer containing chemical active functionalities (acrylic acid) are mixed, in the reaction chamber, with the vapours of a chemically inert aliphatic chain/aromatic rings monomer, styrene, to achieve a sort of controlled "dilution" of the first in the second.

Thus, this mechanism allows to increase the film stability by lowering the number of carboxylic groups at surface and incrementing the cross-linking degree of the polymer in terms of backbone. Due to the higher reticulation, the film becomes more suitable for biological applications where the sample has to be in contact with aqueous biological fluids for a long time. Some preliminary results were obtained [4] from ellipsometric and electrokinetics measurements. For the former, the high T_{Dep} copolymerized coating shows a thickness that is the double of that evaluated for the PPAA coating, obtained with the same discharge parameters and duty cycle. For the latter, the isoelectric point of the thick film (T_{Dep}=7.5 min) obtained in the same operating conditions of PPAA–ST-1 (**Table 10**) was close to pH = 4, typical of inert surfaces.

Four kinds of processes (three in MW and one in CW mode), performed at the same T_{Dep} (5 min), are here below reported and characterized by means of OCA, ATR-FTIR and XPS analysis, in accordance to the parameters optimized for PPAA and PPST and by taking advantage from the previous preliminary copolymerization attempts.

Process type	P_{fwd} (W)	DC	t_{on} (ms)	t_{off} (ms)	P_{AA} (mTorr)	P_{STY} (mTorr)	P_{pro} (mTorr)	Ar_{AA} (sccm)	Ar_{STY} (sccm)
PPAA–ST_1	200	0,1	10	90	25	3	141–150	10	10
PPAA–ST_2	200	0,5	50	50	24	4	141–154	10	10
PPAA–ST_3	200	0,1	20	180	26	4	139–149	10	10
PPAA–ST_4	50	\	\	\	25	4	139–150	10	10

Table 10. Process parameters for plasma copolymerization of acrylic acid and styrene.

Concerning OCA characterization, OCA results of PPAA–PPST films deposited on Corning and silicon substrates are reported, respectively, in **Tables 11** and **12**.

Process type	OCA_{H_2O} (deg)	$OCA_{CH_2I_2}$ (deg)	W_{sl} (mN/m)	W^d_{sl} (mN/m)	W^h_{sl} (mN/m)
PPAA–ST_1	63 ± 2	28.6 ± 0.5	49.15	37.12	12.03
PPAA–ST_2	65 ± 0.5	28.1 ± 0.8	48.37	38	10.37
PPAA–ST_3	18.3 ± 0.7	32.9 ± 0.6	69.98	28.10	41.87
PPAA–ST_4	70.2 ± 0.9	16 ± 2	49.92	43.68	6.25

Table 11. OCA measurements performed on Corning glass substrates coated with PPAA–ST films obtained by plasma-copolymerization processes.

Process type	OCA_{H_2O} (deg)	$OCA_{CH_2I_2}$ (deg)	W_{sl} (mN/m)	W^d_{sl} (mN/m)	W^h_{sl} (mN/m)
PPAA–ST_1	63 ± 1	26.3 ± 0.8	49.7	38.17	11.53
PPAA–ST_2	66.4 ± 0.3	29 ± 2	47.77	38.16	9.61
PPAA–ST_3	15.3 ± 0.7	30.4 ± 0.2	71.14	28.92	42.21
PPAA–ST_4	66 ± 1	25.5 ± 0.6	48.79	39.29	9.50

Table 12. OCA measurements performed on Si substrates coated with PPAA–ST films obtained by plasma-copolymerization processes.

OCA values highlights an hydrophobic behaviour: by comparing OCA results of PPAA–ST copolymers with results of **Tables 2** and **3**, related to PPAA on silicon and Corning glass, respectively, it is clear that OCA_{H2O} values of PPAA–ST films are higher owing to the presence of the styrene component. The only exception is PPAA–ST_3 process. In **Figure 10** are shown the images of main OCA_{H_2O} contact angle related to PPAA–ST_1, PPAA–ST_2 and PPAA–ST_3 processes of **Table 10**.

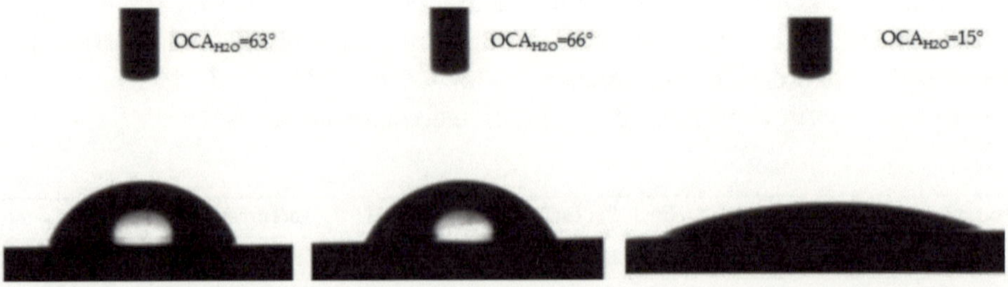

Figure 10. OCA_{H_2O} images for the first three processes listed in **Table 12**, on Si.

FTIR-ATR spectra related to processes listed in **Table 10** are reported in **Figure 11**.

As observed in **Figure 11**, spectra related to plasma copolymers reveal a sum of the features of both the reference polymers (PPAA and PPST). In particular, the spectrum of the film obtained with a t_{on}/t_{off} = 50/50 show characteristics similar to pure PPST. The others, except for the lower intensity of the signals related to the O–H and C=O stretching vibrations, show all the spectral features of carboxylic functionalities of pure PPAA.

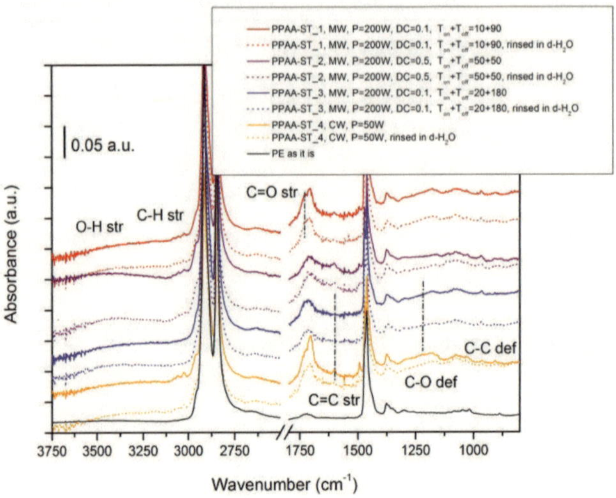

Figure 11. ATR FT-IR spectra measured on PE substrates coated with PPAA–ST copolymerized films, obtained by 7.5 min of deposition.

Under water rinsing, the most show good chemical resistance (except PPAA–ST_4). Under acetone rinsing, only the t_{on}/t_{off} = 50/50 process (PPAA–ST_2) is not dissolved (not shown).

Finally, XPS analysis was carried out in order to evaluate the chemical nature of the resulting copolymers (PPAA–ST): to get the evidence of both benzene rings/aliphatic chain presence and carboxylic functionalities at different grade of ratio. The analysis has been performed on the MW-obtained coatings only, listed in **Table 10**.

Figure 12 reports the C1s HR spectrum (a) and the O1s HR spectrum (b) of process PPAA–ST_2. Peaks have been deconvoluted using Gaussian–Lorentz functions, while backgrounds have been removed with Tougaard functions. Fitting peaks are:

- The lower energy peak (284.3 eV) has been assigned to ring carbons,
- The peak at 284.5 eV has been attributed to the presence of C atoms linked as C–C/C–H,
- The peak at 285.9 eV is commonly attributed to C–O–C/C–O–H,
- The peak at 288.6 is commonly attributed to COOH, O–C=O,
- The higher energy peak (~291 eV) is due to π–π^* transitions arising from the presence of the aromatic rings,
- The peak at 532.2 eV is attributed to C–O–C,
- The peak 533.5 eV is commonly attributed to COOH/C=O.

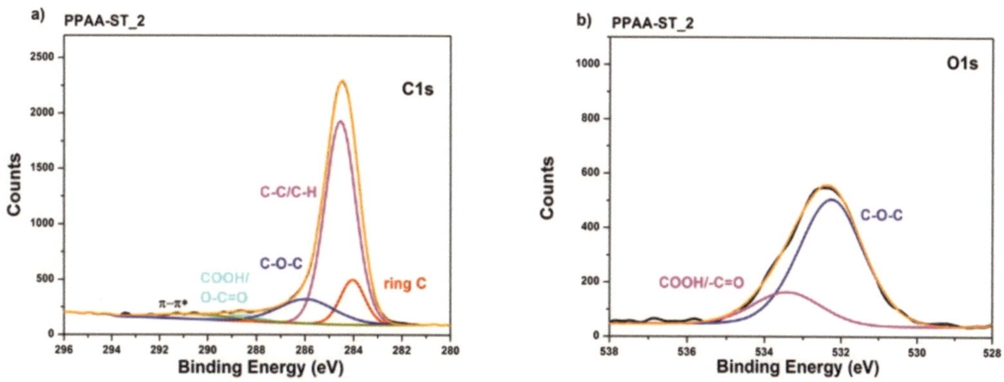

Figure 12. HR C1s spectrum (a) and HR O1s spectrum (b) of PPAA–ST_2 process (DC = 0.5 and t_{on}/t_{off} ratio = 50/50).

For process PPAA–ST_2, the C1s component related to π–π^* transition is higher than in spectrum related to process PPAA–ST_3 performed by a different DC and t_{on}/t_{off} ratio (0.1 and 20/180). In this last (**Figure 13**), all the components, of both HR C1s and HR O1s, are related to the higher embedding of acrylic acid species during chain propagation.

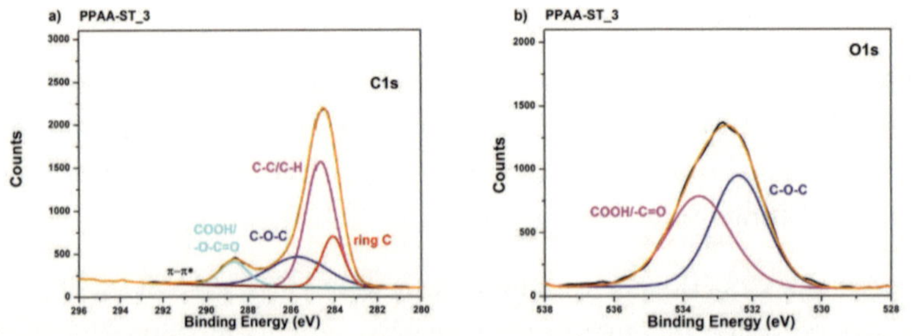

Figure 13. HR C1s spectrum (a) and HR O1s spectrum (b) of PPAA–ST process (PPAA–ST_3) at P = 200W; DC = 0.1; 20ms/180ms.

The comparison of all HR-C1s spectra (**Figure 14**) shows that the content of carbon and related components are different each other, as function of processes parameters. In particular, after rinsing of PPAA–ST_2 sample in acetone (spectrum PPAA–ST_2I), a little part of π–π^* transition component is lost and the carboxylic component increases, but this is the only kind of tested coating that quite resists to acetone treatment (in view of the photolithographic process for lateral patterning).

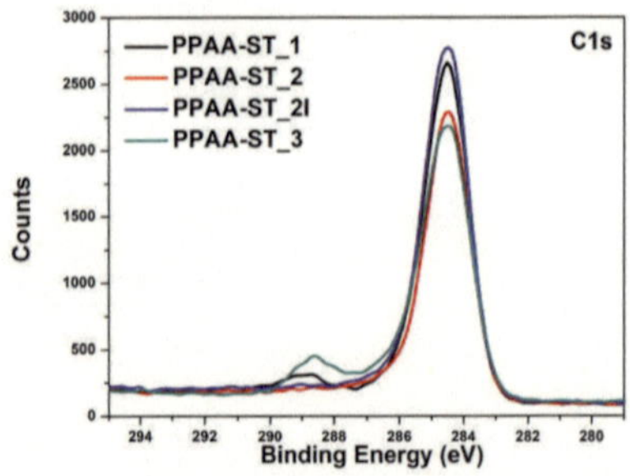

Figure 14. Comparison of HR C1s spectra for the first three processes listed in **Table 10.**

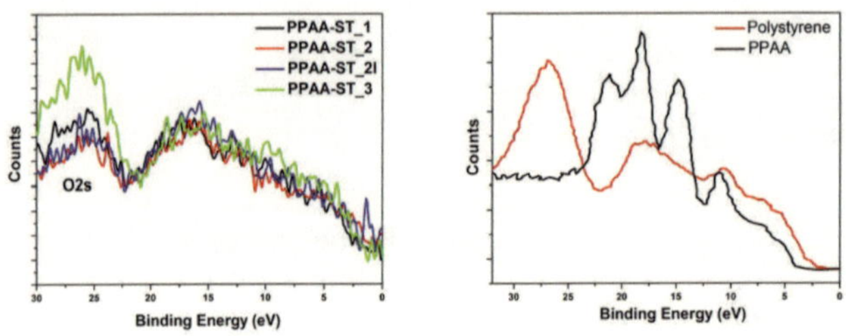

Figure 15. Comparison between all PPAA–ST processes valence band (left) HR spectra and reference valence bands spectra of polystyrene and a plasma-polymerized acrylic acid, reported in literature [32, 33] (right-image re-plotted by authors).

Notwithstanding evident differences among the spectra, all of them are quite consistent, in the low energy range of valence band, with the literature reference for plasma-polymerized acrylic acid, as shown in **Figure 15** (red spectrum in the graph on the left).

The trend of O/C ratio (**Table 13**) confirms that the process, producing a higher amount of oxygen-containing species (–COOH groups), is the PPAA–ST_3 and the most PPST-like one is the PP PPAA–ST_2 process.

Process type	Dep time	PPST films	O/C atomic percentage ratio
PPAA-ST_1	7.5 min	P = 200W; DC. 0.1; 10ms/90ms	0.16
PPAA-ST_2	7.5 min	P = 200W; DC. 0.1; 50ms/50ms	0.11
PPAA-ST_2I	7.5 min	P = 200W; DC. 0.1; 50ms/50ms + 5′ in acetone	0.10
PPAA-ST_3	7.5 min	P = 200W; DC. 0.1; 20ms/180ms	0.25

Table 13. O/C atomic ratio referred to elements percentage obtained from survey spectra of PPAA–ST films.

A colorimetric titration with TBO has been performed on PPAA–ST coatings deposited on Si according to different process parameters, after H_2O rinsing. Through this approach, it is possible to quantitatively estimate the surface density of –COOH groups exposed at the film surface and available for biomolecules grafting.

Table 14 reports surface density data obtained for several processes performed in CW and MW modes, on Si substrates. The results match the outcomes of previous characterizations: for DC = 0.5, the content of carboxylic reacting groups is under technique limit of detection, while for DC = 0.1 and CW process, the content is comparable with PPAA carboxylic groups density and so promising for biomolecules grafting.

Process type	Processes parameters	N°COOH/cm²
PPAA–ST_1	Si-MW P_{ave} = 20 W; DC = 10%; t_{on} = 10 ms, t_{off} = 90 ms; Ar Etch	~5 × 10^{15}
PP AA-ST_2	Si-MW P_{ave} = 4 W; DC =50%; t_{on} = 50 ms, T_{off} = 50 ms; Ar Etch	0
PP AA-ST_3	Si-MW P_{ave} = 20 W; DC = 10%; t_{on} = 20 ms, t_{off} = 180ms; Ar Etch	~3 × 10^{15}
PP AA-ST_4	Si-50 W; Ar Etch	~4 × 10^{15}

Table 14. Carboxylic group density for plasma processes of Table 10.

In order to get a calibration of thicknesses of PPAA–ST in the lower range possible, already reported for PPAA (10–50 nm) according to deposition time, AFM was used with a profilometric approach.

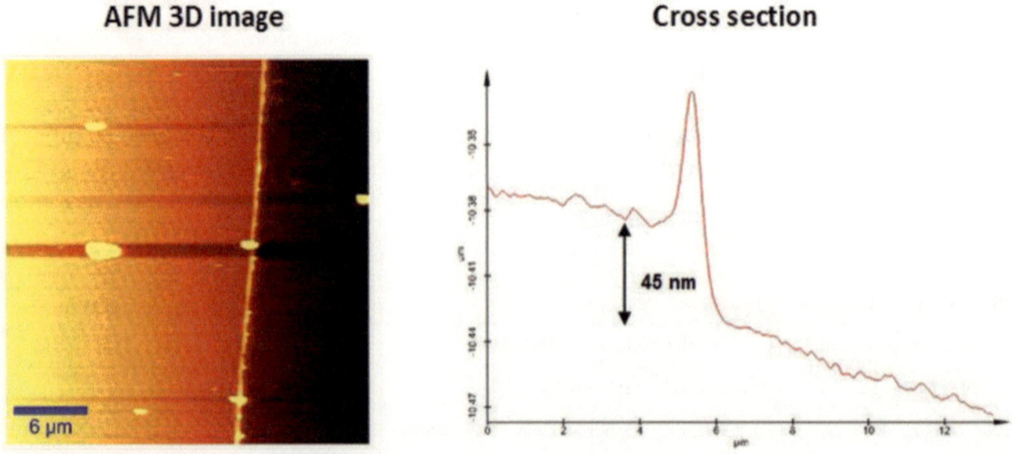

Figure 16. PPAA–ST_2 (T_{dep} =30 s) AFM characterization.

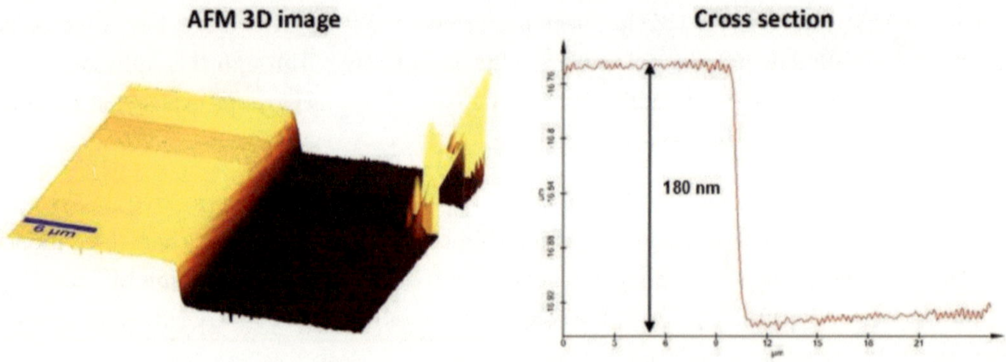

Figure 17. PPAA–ST_2 (T_{dep} = 5 min) AFM characterization.

PPAA–ST_2: DC 50% (t_{on} 50 ms t_{off} 50 ms) P = 200 W				
Time dep (min)	**Time dep (s)**	**Thickness (nm)**	**Dep rate (nm/s)**	**Dep rate (nm/min)**
0.5 (Figure 16)	30.0	45.0	1.50	90
1	60.0	80.0	1.33	80
5 (Figure 17)	300.0	180.0	0.60	36

Table 15. AFM measure: PPAA–PPST thickness values at different deposition time.

From **Table 15** arises that the deposition rate has not a linear behaviour and is dramatically huge with respect the other plasma polymers. The same behaviour is observed for the other processes of **Table 10**.

http://dx.doi.org/10.5772/62899

For PPAA–ST, the minimum deposition time possible, providing about 40-nm thick coatings, is 30 s, but this is quite uncontrollable due to the starting step of stabilization occurring when pulsed plasma discharge is activated.

4. Conclusions

In this work, the reliability to obtain, through a methodic approach, ultra-thin plasma polymerized coatings, about 40-nm thick, able to retain the chemical properties of the pristine precursor, for PPAA and PPST, in terms of binding and bio-antifouling, respectively, notwithstanding the quite low thickness and thus the short extension of macromolecular crosslinking in Z direction. PPAA coatings expose accessible carboxylic groups at a different grade of surface density (up to about 10^{16} –COOH groups/cm^2) and are stable in several aqueous media. The most performing PPST coating, with hydrophobic and so antifouling properties (that discourage proteins adsorption), resists also to solvent immersion, such as acetone.

Finally, an ultra-thin plasma copolymer, PPAA–ST, that merges the properties of PPAA and PPST was developed in order to enhance chemical stability of the chemically functional coating by diluting hydrophilic carboxylic functionalities. Moreover, the introduction of a higher content of aliphatic/aromatic components provided the reinforcement of the coating carbon backbone.

It is worth of note that, for a particular AA/ST ratio (PPAA–ST_1 process with DC = 10%; t_{on} = 10 ms, t_{off} = 90 ms), notwithstanding the increase in aliphatic fragments/aromatic rings content and thus in chemical resistance, the number of carboxylic groups exposed at the surface was shown to be only one order of magnitude lower than the one obtained for PPAA.

Acknowledgements

This research has received funding from the project BILOBA (no. 318035) granted in the frame of EU FP7/2007–2013 Programme. The authors thank Dr Mirko Nietschke (Institute für Polymer Vorschung, Dresden, Germany) for the support for the ellipsometric and electrokinetics measurements.

Author details

Paola Rivolo[1*], Micaela Castellino[2], Francesca Frascella[1] and Serena Ricciardi[1]

*Address all correspondence to: paola.rivolo@polito.it

1 Department of Applied Science and Technology, Politecnico di Torino, Turin, Italy

2 Centre for Space Human Robotics, Istituto Italiano di Tecnologia, Turin, Italy

References

[1] Muguruma H. Plasma polymerized films for biochips design. Plasma Processes and Polymers. 2010;7(2):151–162. DOI: 10.1002/ppap200900125

[2] Hessner M.J., Meyer L., Tackes J., Muheisen S., Wang X. Immobilized probe and glass surface chemistry as variables in microarray fabrication. BMC Genomics. 2004;5(1):53–61. DOI: 10.1186/1471-2164-5-53

[3] Cretich M., Damin F., Pirri G., Chiari M. Protein and peptide arrays: Recent trends and new directions. Biomolecular Engineering. 2006;23(2–3):77–88. DOI: 10.1016/j.bioeng.2006.02.001

[4] Ricciardi S. Surface chemical functionalization based on plasma techniques. Germany: LAP Lambert Academic Publishing; 2012. 236 p. ISBN-13: 978–3659190117

[5] Zhang Z., Menges B., Timmons R.B., Knoll W., Förch R. Surface plasmon resonance studies of protein binding on plasma polymerized di(ethylene glycol) monovinyl ether films. Langmuir. 2003;19(11):4765–4770. DOI: 10.1021/la026980d

[6] Liu S., Vareiro M.M., Fraser S., Jenkins A.T. Control of attachment of bovine serum albumin to pulse plasma-polymerized maleic anhydride by variation of pulse conditions. Langmuir. 2005;21(19):8572–8575. DOI: 10.1021/la051449e

[7] Detomaso L., Gristina R., Senesi G.S., d'Agostino R., Favia P. Stable plasma-deposited acrylic acid surfaces for cell culture applications. Biomaterials. 2005;26(18):3831–3841. DOI: 10.1016/j.biomaterials.2004.10.011

[8] Mourtas S., Kastellorizios M., Klepetsanis P., Farsari E., Amanatides E., Mataras D., et al. Covalent immobilization of liposomes on plasma functionalized metallic surfaces. Colloids Surface B: Biointerfaces. 2011;84(1):214–220. DOI: 10.1016/j.colsurfb.2011.01.002

[9] Kizling M., Järås S.G. A review of the use of plasma techniques in catalyst preparation and catalytic reactions. Applied Catalysis A: General. 1996;147(1):1–21. DOI: 10.1016/S0926-860X(96)00215-3

[10] Denes F.S., Manolache S. Macromolecular plasma-chemistry: An emerging field of polymer science. Progress in Polymer Science. 2004;29(8):815–885. DOI: 10.1016/j.progpolymsci.2004.05.001

[11] Förch R., Chifen A.N., Bousquet A., Khor H.L., Jungblut M., Chu L.-Q., et al. Recent and expected roles of plasma-polymerized films for biomedical applications. Chemical Vapour Deposition. 2007;13(6–7):284–294. DOI: 10.1002/cvde.200604035

[12] Rinsch C.L., Chen X., Panchalingam V., Eberhart R.C., Wang J.-H., Timmons R.B. Pulsed radio frequency plasma polymerization of allyl alcohol: Controlled deposition of surface hydroxyl groups. Langmuir. 1996;12(12):2995–3002. DOI: 10.1021/la950685u

[13] Yasuda H., Hsu T. Some aspects of plasma polymerization investigated by pulsed R.F. discharge. Journal of Polymer Science: Polymer Chemistry Edition. 1977;15(1):81–97. DOI: 10.1002/pol.1977.170150109

[14] Jafari R., Tatoulian M., Morscheidt W., Arefi-Khonsari F. Stable plasma polymerized acrylic acid coating deposited on polyethylene (PE) films in a low frequency discharge (70 kHz). Reactive and Functional Polymers. 2006;66(12):1757–1765. DOI: 10.1016/j.reactfunctpolym.2006.08.006

[15] Finke B., Schröder K., Ohl A. Structure retention and water stability of microwave plasma polymerized films from allylamine and acrylic acid. Plasma Processes and Polymers. 2009;6(1):S70–S74. DOI: 10.1002/ppap.200930305

[16] Morent R., De Geyter N., Trentesaux M., Gengembre L., Dubruel P., Leys C., et al. Stability study of polyacrylic acid films plasma-polymerized on polypropylene substrates at medium pressure. Applied Surface Science. 2010;257(2):372–380. DOI: 10.1016/j.apsusc.2010.06.080

[17] Lin Y., Yasuda H. Hydrocarbon barrier performance of plasma-surface-modified polyethylene. Journal of Applied Polymer Science. 1996;60(12):2227–2238. DOI: 10.1002/(SICI)1097-4628(19960620)60:12<2227::AID-APP21>3.0.CO;2–2

[18] Kelly J.M., Short R.D., Alexander M.R. Experimental evidence of a relationship between monomer plasma residence time and carboxyl group retention in acrylic acid plasma polymers. Polymer. 2003;44(11):3173–3176. DOI: 10.1016/S0032-3861(03)00217-9

[19] Anders A. Fundamentals of pulsed plasmas for materials processing. Surface & Coatings Technology. 2004;183(2–3):301–311. DOI: 10.1016/j.surfcoat.2003.09.049

[20] Swaraj S., Oran U., Friedrich J.F., Lippitz A., Unger W.E.S. Surface chemical analysis of plasma-deposited copolymer films prepared from feed gas mixtures of ethylene or styrene with allyl alcohol. Plasma Processes and Polymers. 2007;4(4):376–389. DOI: 10.1002/ppap.200600215

[21] Sardella E., Gristina R., Ceccone G., Gilliland D., Papadopoulou-Bouraoui A., Rossi F., et al. Control of cell adhesion and spreading by spatial microarranged PEO-like and pdAA domains. Surface & Coatings Technology. 2005;200(1–4):51–57. DOI: 10.1016/j.surfcoat.2005.02.063

[22] Ballarini M., Frascella F., De Leo N., Ricciardi S., Rivolo P., Mandracci P., et al. A polymer-based functional pattern on one-dimensional photonic crystals for photon sorting of fluorescence radiation. Optics Express. 2012;20(6):6703–6711. DOI: 10.1364/OE.20.006703

[23] Ricciardi C., Ferrante I., Castagna R., Frascella F., Marasso S.L., Santoro K., et al. Immunodetection of 17deestradiol in serum at ppt level by microcantilever resonators. Biosensors and Bioelectronics. 2013;40(1):407–411. DOI: 10.1016/j.bios.2012.08.043

[24] Ballarini M., Frascella F., Michelotti F., Digregorio G., Rivolo P., Paeder V., et al. Bloch surface waves-controlled emission of organic dyes grafted on a one-dimensional

photonic crystal. Applied Physics Letters. 2011;99(4):043302–043305. DOI: 10.1063/1.3616144

[25] Ballarini M., Frascella F., Enrico E., Mandracci P., De Leo N., Michelotti F., et al. Bloch surface waves-controlled fluorescence emission: Coupling into nanometer-sized polymeric waveguides. Applied Physics Letters. 2012;100(6):063305–063309. DOI: 10.1063/1.3684272

[26] Frascella F., Ricciardi S., Rivolo P., Moi V., Giorgis F., Descrovi M., et al. A fluorescent one-dimensional photonic crystal for label-free biosensing based on Bloch surface waves. Sensors. 2013;13(2):2011–2022. DOI: 10.3390/s130202011

[27] Descrovi E., Morrone D., Angelini A., Frascella F., Ricciardi S., Rivolo P., et al. Fluorescence imaging assisted by surface modes on dielectric multilayers. The European Physical Journal E. 2014;68(3):53–55. DOI: 10.1140/epjd/e2014-40530-0

[28] Chen J.-P., Chiang Y.-P. Surface modification of non-woven fabric by DC pulsed plasma treatment and graft polymerization with acrylic acid. Journal of Membrane Science. 2006;270(1–2):212–220. DOI: 10.1016/j.memsci.2005.11.015

[29] Blanchemain N., Aguilar M.R., Chai F., Jimenez M., Jean-Baptiste E., El-Achari A., et al. Selective biological response of human pulmonary microvascular endothelial cells and human pulmonary artery smooth muscle cells on cold-plasma-modified polyester vascular prostheses. Biomedical Materials. 2011;6(6):065003–065014. DOI: 10.1088/1748-6041/6/6/065003

[30] Ricciardi S., Castagna R., Severino S.M., Ferrante I., Frascella F., Rivolo P., et al. Surface functionalization by poly-acrylic acid plasma-polymerized films for microarray DNA diagnostics. Surface & Coatings Technology. 2012;207:389–399. DOI: 10.1016/j.surfcoat.2012.07.026

[31] Rivolo P., Severino S.M., Ricciardi S., Frascella F., Geobaldo F. Protein immobilization on nanoporous silicon functionalized by RF activated plasma polymerization of acrylic acid. Journal of Colloid and Interface Science. 2014;416:63–80. DOI: 10.1016/j.jcis.2013.10.060

[32] Way W.K., Rosencrance S.W., Winograd N., Shirley D.A. Polystyrene by XPS. Surf Science Spectra. 1993;2:67–70. DOI: 10.1116/1.1247712

[33] O'Toole L., Beck A.J., Short R.D. Characterization of plasma polymers of acrylic acid and propanoic acid. Macromolecules. 1996;29:5172–5177. DOI: 10.1021/ma9518417

Chapter 6

Amorphous Silicon Photonics

Ryohei Takei

Additional information is available at the end of the chapter

http://dx.doi.org/10.5772/63374

Abstract

This chapter introduces our research on amorphous silicon photonics. By exploring our high-quality silicon thin-film technology, we have demonstrated hydrogenated amorphous silicon (a-Si:H) waveguides with ultra-low-loss, vertical interlayer transition (VIT) devices for cross coupling between vertically stacked optical circuits. These device technologies are promising for three-dimensional photonic integrated circuits integrated in microelectronics chips. A record low loss of 0.6 dB cm^{-1} was achieved for a submicron-scale single-mode waveguide, and the VIT devices allow low-loss, broadband, and polarization-insensitive operation.

Keywords: hydrogenated amorphous silicon, silicon photonics, 3D integration, photonic integrated circuit, backend integration

1. Introduction

Amorphous silicon (a-Si) is a well-known and industrially proven non-crystalline material. It is widely used in commercial applications; for example, thin-film transistors in liquid crystal displays, thin-film solar cells, and microbolometers for thermal cameras [1]. Recently, the mature silicon (Si) thin-film technology has advanced toward a new research field: Si photonics. Si photonics is an ultrahigh-density integration technology for optical devices using submicron-scale optical waveguides with a core composed of Si, which is mainly attractive for data communication applications. Typically, optical Si-integrated devices are fabricated on silicon-on-insulator (SOI) wafers consisting of crystalline Si (c-Si) and buried oxide (BOX). The large difference in refractive index (n) between Si (n = ~3.5) and silica (SiO_2) (n = ~1.4) enables miniaturization of such optical devices. Meanwhile, Si thin films have received attention because of their favorable deposition and optical properties. Through deposition of Si, three-dimensional (3D) photonic integrated circuits (PICs) can be constructed by vertical stacking of multiple a-Si waveguide layers clad with an amorphous insulating material such as SiO_2. Such vertical stacking would be inherently infeasible on a c-Si platform because it is virtually impossible to deposit c-Si on an amorphous substrate such as thermal oxide. Furthermore, a-Si has higher optical nonlinearity and lower nonlinear absorption than the corresponding values of c-Si. Thus, a-Si is an attractive material for use in optical nonlinear devices such as parametric amplifiers, all-optical signal processing devices, and all-optical modulators [2–9].

A disadvantage of a-Si is that it has an optical absorption loss in the infrared wavelength range assigned for optical telecommunication. In a-Si, Si atoms form a random network and have dangling bonds that cause absorption loss of infrared light. This disadvantage has been overcome by passivation of the dangling bonds by hydrogen. Two decades ago, Cocorullo et al. [10] reported the first example of hydrogenated a-Si (a-Si:H) waveguides. They deposited a-Si:H films through plasma decomposition of silane (SiH_4). In 2005, Harke and colleagues developed a single-mode a-Si:H waveguide on a thermal oxide with a propagation loss of 2.0 dB cm^{-1} at a wavelength of 1.55 μm for the ridge waveguide with a height of 1.3 μm and a width of 1.1 μm [11]. Researchers at Ghent university demonstrated an a-Si:H nanophotonic waveguide using a complementary metal-oxide semiconductor (CMOS)-compatible process [12]. A propagation loss of 3.5 dB cm^{-1} was achieved for the wire waveguide with a width of 480 nm and height of 220 nm. The optimized KrF lithography and Si dry-etching process decreased the roughness at the waveguide sidewalls, resulting in a low-loss wire waveguide. Also, Zhu et al. [13] from the Institute of Microelectronics reported a-Si:H wire waveguides fabricated using a CMOS-compatible process. The propagation loss of the waveguides was 3.2 dB cm^{-1}, and they were 200 nm high and 500 nm wide. Chemical mechanical polishing (CMP) was used to smooth the a-Si:H top surface. Sun et al. [14] from the Massachusetts Institute of Technology achieved a propagation loss of 2.7 dB cm^{-1} for waveguides with a width of 700 nm and height of 100 nm with a thin interlayer cladding made of silicon nitride (SiN). SiN has n = ~2.0 which is between those of Si and SiO_2. Thus, the SiN interlayer mitigates the difference of n between Si and SiO_2, resulting in a decreased propagation loss. Kang et al. [15] from Tokyo Institute of Technology reported vertically stacked a-Si:H waveguides with two waveguide layers on an SOI waveguide layer by repeated deposition of a-Si:H and SiO_2.

We have focused on a-Si:H 3D PIC technology, which is promising to replace electrical interconnect in CMOS chips with optical interconnects. Electrical interconnects are currently one of the bottlenecks for the performance improvement of CMOS chips because the aggregate bandwidth of intra-chip data transmission is physically limited by power consumption and the frequency of the clock signal. In contrast, optical interconnects could provide higher speed and larger bandwidth data transmission than electrical ones because the frequency of light as a data carrier is much higher than that of electricity and wavelength division multiplexing is available. Furthermore, a-Si:H photonics has two additional advantages: potential feasibility of 3D PICs and process compatibility with CMOS backend processes. Regarding the former, 3D PIC could increase the bandwidth by stacking of multiple optical layers. Meanwhile, integration of optical interconnection with CMOS chips in a back-end-of-line configuration is possible because the deposition temperature of a-Si:H is less than 400°C. In this chapter, we summarize our research achievements. In Section 2, we report a-Si:H waveguides with ultra-low loss together with high-quality a-Si:H technology, which is fundamental for high-performance PICs. In Section 3, the concept of vertical interlayer transition (VIT) devices is described. The optimal design, fabrication, and measurement results of our VIT devices are presented.

2. Low-loss a-Si:H waveguide

2.1. High-quality a-Si:H

Because a-Si has an absorption loss in the telecommunication wavelength range, development of low-loss a-Si:H is important. High-quality a-Si:H films can be obtained by plasma-enhanced chemical vapor deposition (PECVD) through plasma decomposition of a source gas mixture of SiH_4 and hydrogen (H_2). The deposition conditions strongly influence the characteristics of the deposited film. Deposition temperature is the most important factor affecting film quality. On the growth surface during the deposition, the main precursor (SiH_3 radicals) reaches the surface and then diffuses on it to find energetically stable sites. As the deposition temperature increases, the diffusion of the precursor is enhanced. As a result, the defect density is lowered. However, desorption of the hydrogen covering the surface occurs above 250°C, which increases the defect density. Thus, the defect density is determined by the balance of these two reactions: diffusion and desorption. The lowest defect density is achieved in a-Si:H films deposited at 250°C [16].

We measured the sub-gap absorption coefficient of a 1-μm-thick a-Si:H film deposited at 250°C [17]. A highly sensitive measurement technique is required to do this because a well-passivated a-Si:H film has low absorption. The absorption of the film was measured using the constant photocurrent method (CPM) calibrated with the transmittance and reflection (T&R) spectra. CPM is a highly sensitive but relative measurement, whereas T&R spectra are absolute measurements but have low sensitivity. In the high-absorption region, the absorption coefficient of a material can be determined from T&R spectra. In contrast, CPM can measure the relative absorption over the whole wavelength region including those with high and low absorption. The absorption coefficient in the low-absorption region can be determined by calibrating the CPM spectra with the T&R measurements in the high-absorption region. The measured absorption coefficient in **Figure 1** was 10^{-2} cm^{-1} at a 1550 nm, corresponding to a loss of 0.04 dB cm^{-1}. The defect density of the film was 4.2×10^{15} cm^{-3}, which was evaluated from the integrated defect absorption using a conversion factor of 1.9×10^{16} cm^{-2} eV^{-1} [18]. **Table 1** compares the material losses of a-Si:H reported to date, showing that the loss of our a-Si:H film is the lowest obtained so far.

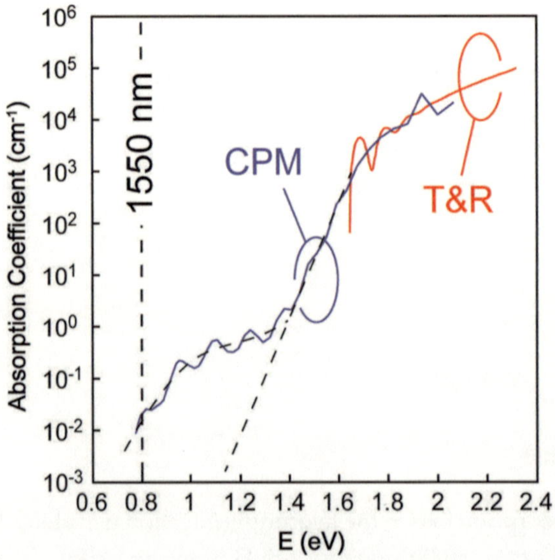

Figure 1. Sub-gap absorption spectrum of a 1-μm-thick a-Si:H film measured using the CPM and T&R approach.

Reference	Deposition temp. (°C)	Material loss (dB/cm)	Method
[19]	Not stated	< 1	Estimation from waveguide loss dependence on width
[20]	Not stated	0.5	Prism coupling method
[12]	300	0.7	Estimation from waveguide loss
[21]	220	1.1	Photothermal deflection spectroscopy
Our work [17]	250	0.04	CPM calibrated with T&R spectra

Table 1. Comparison of material losses in a-Si:H films.

2.2. Waveguide fabrication

Nanophotonic waveguides with a core made of our low-absorption a-Si:H films were fabricated and their propagation losses were evaluated [17]. Two types of waveguides were tested: one composed of 420-nm-wide wires and another with 780-nm-wide ridges. For reference, c-Si waveguides with the same geometry were also tested. The waveguides were fabricated as follows. First, 220-nm-thick a-Si:H films were deposited by PECVD using a gas mixture of SiH_4 and H_2 at 250°C on a Si wafer with a 2-μm-thick thermal oxide layer. The waveguides were formed using a standard microfabrication technology including i-line lithography. The a-Si:H film used for the wires was dry etched through inductively coupled plasma (ICP)-reactive ion etching (RIE) using a gas mixture of sulfur hexafluoride (SF_6) and octafluorocyclobutane (C_4F_8). In contrast, the a-Si:H film used for the ridges was etched by carbon tetrafluoride-based capacitively coupled plasma-RIE. After removing the remaining photore-

http://dx.doi.org/10.5772/63374

sist and by using a wafer-cleaning process, the 1.5-μm-thick upper cladding of the waveguides was deposited by plasma decomposition of a gas mixture of tetraethyl orthosilicate and oxygen at 350°C. The reference c-Si waveguides were fabricated following a similar process on an SOI wafer composed of a 220-nm-thick top Si layer and underlying 2-μm-thick BOX layer. Finally, the wafers were cleaved to form smooth edge facets for light coupling from/to optical fibers. **Figure 2** depicts cross-sectional scanning electron microscope (SEM) images of the fabricated waveguides. The main electrical field profiles of the quasi-transverse electric (q-TE) mode are also shown. The field profiles were numerically calculated by the finite difference method (FDM). Although both c-Si and a-Si:H were etched under the same conditions, the etching depth of a-Si:H was slightly deeper than that of c-Si. This indicates that the etching rate of a-Si:H may be slightly higher than that of c-Si. The difference in the etching depths did not affect the waveguide performance.

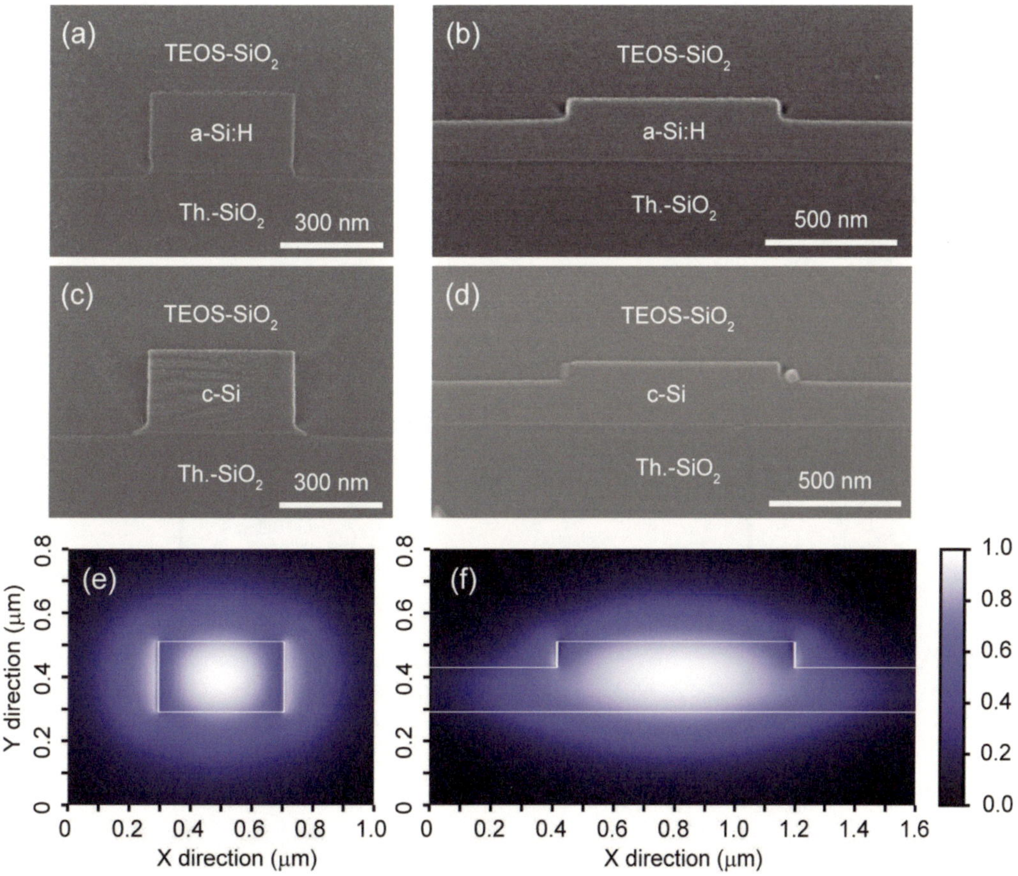

Figure 2. (a)–(d) Cross-sectional SEM images of the fabricated wire and ridge waveguides. (a) a-Si:H wire, (b) a-Si:H ridge, (c) c-Si wire, and (d) c-Si ridge waveguides. (e)–(f) Normalized main electrical field profiles calculated by the finite difference method.

2.3. Results and discussion

The propagation losses of the fabricated waveguides were evaluated using a standard cutback method. To laterally couple light from/to all waveguides through the common edge of a device chip, waveguides were arranged in a convoluted manner with a curve radius of 10 μm for the wire waveguides and 400 μm for the ridge waveguides. The device chips contained 12 wire waveguides or 16 ridge waveguides with lengths ranging from 0.5 to 1.7 cm. Numerical analysis based on FDM indicated that the single mode conditions were satisfied for both the wire and ridge waveguides. The bending losses were also estimated to be negligible for the q-TE mode at 1.55 μm. The loss measurement was carried out for the q-TE mode only because the ridge waveguides do not support a quasi-transverse magnetic (q-TM) mode.

Light emitted from an amplified spontaneous emission (ASE) source was filtered by wavelength-tunable filters with 1-nm bandwidth and polarized, resulting in transverse electric (TE)-polarized light around 1.55 μm. The light was directed into the waveguides through a lens-tipped polarization maintaining fiber. The output light was collected using a lens-tipped single-mode fiber to send the output to an optical power meter. **Figure 3** plots the measured transmittance against waveguide length. The propagation losses were extracted from the slopes of the linear fittings. The propagation losses were 4.14 ± 0.15 and 1.20 ± 0.08 dB cm^{-1} for the wire and ridge a-Si:H waveguides, respectively. The SOI waveguides had losses of 2.31 ± 0.21 and 0.66 ± 0.06 dB cm^{-1} for the wire and ridge waveguides, respectively.

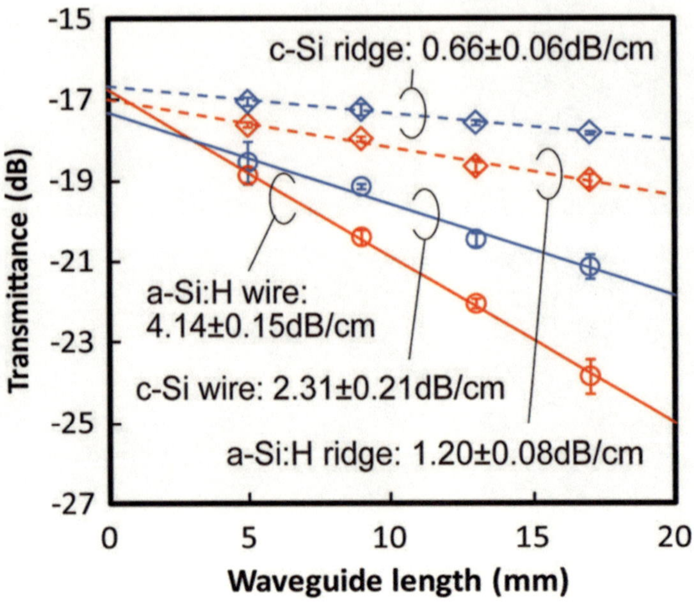

Figure 3. Measured propagation losses of wire and ridge waveguides with 220-nm-thick Si cores. The error bars show the standard deviations (±σ) of the measured transmittances.

To measure the wavelength dependence of the transmittance of the waveguides in the C-band from 1530 to 1565 nm, TE-polarized ASE light was fed into the 17-mm-long waveguides, and the output light was measured using an optical spectrum analyzer. **Figure 4** shows the measured dependences of transmittance on wavelength. The ridge waveguides have smoother curves than the wire waveguides, indicating that the ripples may originate from light scattering by the roughness of the waveguide sidewall. Such roughness is unintentionally formed during the lithography and etching processes.

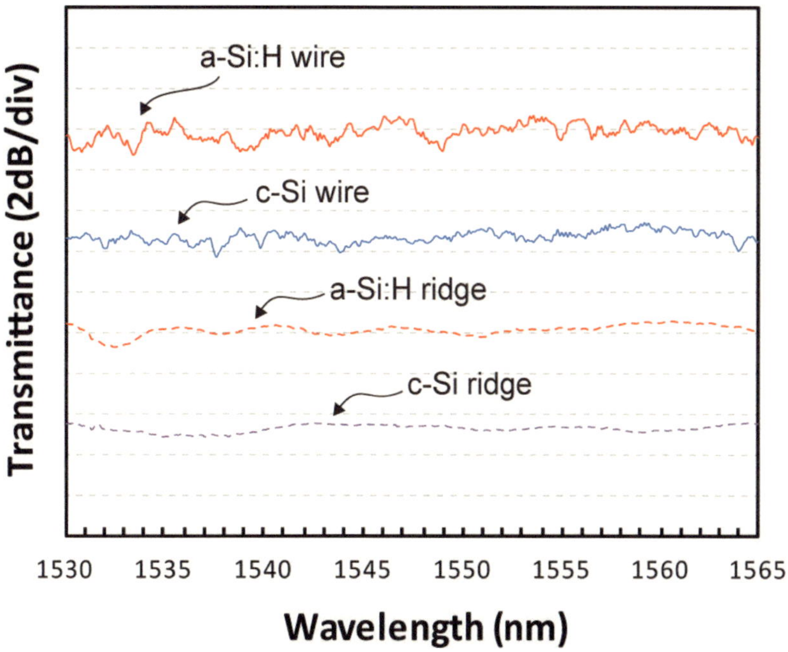

Figure 4. Measured dependences of the transmittance of each fabricated waveguide on wavelength.

Our a-Si:H has a comparable material loss to c-Si. However, the a-Si:H waveguides show higher propagation loss than their c-Si counterparts even though both the a-Si:H and c-Si waveguides were fabricated using the same process. The presence of microvoids in the a-Si:H films is a possible origin of their propagation loss. It has been argued that the dihydride SiH_2 and trihydride SiH_3 make a-Si:H films heterogeneous, producing H clusters or microvoids in the films [22]. Fourier transform infrared absorption spectroscopy of our a-Si:H films revealed a stretching mode from the monohydride SiH around 2000 cm^{-1} (7–8% H content); no shoulder or peak was observed at higher wavenumber. Therefore, the density of microvoids in our a-Si:H films would be sufficiently low that the light scattering at the microvoids is negligible. In contrast, the top surface of the as-grown a-Si:H is approximately three times rougher than that of the c-Si on a SOI wafer, as illustrated in the atomic force microscope (AFM) images in **Figure 5**. The loss differences between the a-Si:H and c-Si waveguides probably originates from the roughness of the top surface, considering that the absorption of the upper and lower cladding materials is negligible [23].

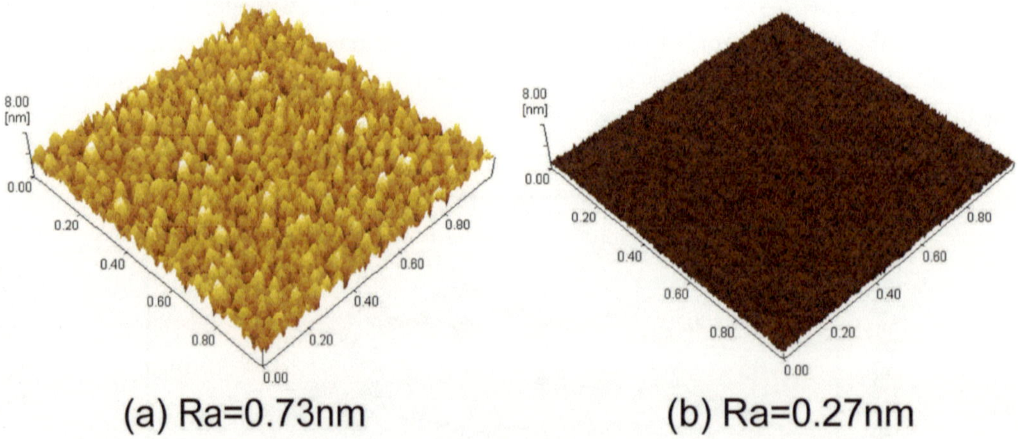

Figure 5. AFM images of the surfaces of (a) as-grown a-Si:H and (b) a commercially available SOI wafer.

Photoinduced degradation is an intrinsic problem of a-Si:H, which is caused by the Si dangling bonds that are generated through non-radiative recombination of photoexcited electron-hole pairs [24]. The degradation of a-Si:H in the telecommunication wavelength range has been reported [5, 9]. In waveguide applications of a-Si:H, the dangling bonds may cause an additional absorption loss. In particular, several research groups are concerned about photo-induced degradation because high-power light is input into the waveguides for optical nonlinear devices. Fortunately, the loss of our a-Si:H waveguides did not increase during measurement. This is probably because our a-Si:H has an optical gap of 1.75 eV, as determined by Tauc plot. Such a wide bandgap markedly decreases two-photon absorption, which suppresses non-radiative recombination of photogenerated carriers and thus photoinduced degradation.

2.4. Waveguide with ultra-low loss

In the previous section, we showed that a-Si:H has a rougher surface than c-Si, resulting in an increased propagation loss. Similarly, the roughness of the waveguide sidewalls contributes to the loss. Thus, minimizing roughness is an effective way to further lower the loss in a-Si:H waveguides. In this work, two approaches to lower roughness were tested. One is a planarization of the top surface of a-Si:H by CMP. The other was introduction of a shallow-ridge geometry. When the material loss of a core is sufficiently low, shallow ridges can lower the propagation loss because of the low depth of the etched sidewalls. First, a 500-nm-thick a-Si:H film was deposited on a 2-μm-thick thermal oxide top layer on a Si wafer. Then, the as-grown surface layer was polished by CMP, resulting in the removal of approximately 60 nm of the a-Si:H layer. **Figure 6(a)** shows the polished surface observed by AFM, which has a roughness comparable to that of a c-Si surface. Ridge waveguides with a ridge height of 100 nm were then formed on the polished a-Si:H, as depicted in **Figure 6(b)**. **Figure 7** presents the measurement results for the a-Si:H shallow-ridge waveguide. The measured propagation loss was 0.60 ± 0.02 dB cm^{-1} for the q-TE mode at 1.55 μm, which is the lowest reported propagation loss for an a-Si:H waveguide. Also, the dependence of the transmittance on the wavelength in the C-band is within a 0.5-dB range for the 17-mm-long waveguide (inset of **Figure 7**). Zhu et al. [13] reported that the CMP process may generate new Si dangling bonds at the surface, substantially increasing the propagation losses. While our CMP process lowered the losses, the passivation of surface defects may be needed to further decrease propagation losses.

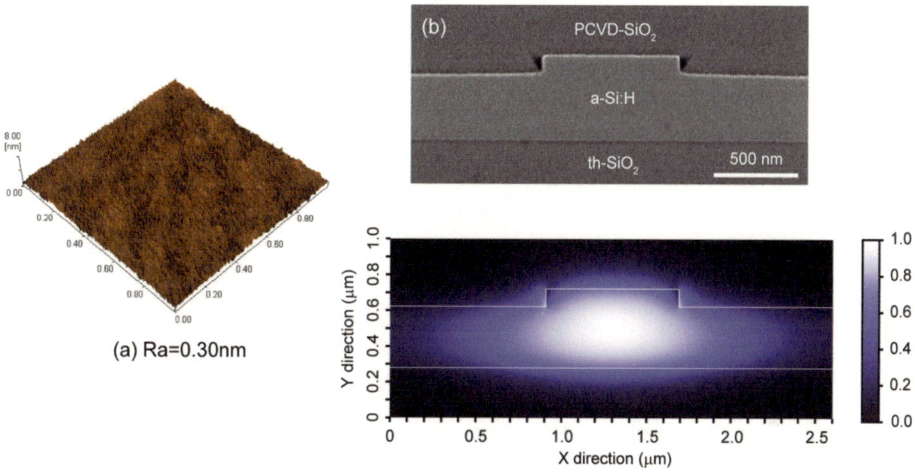

Figure 6. (a) AFM image of a polished a-Si:H surface, (b) cross-sectional SEM image, and (c) main electrical mode profile of the q-TE mode for a shallow-ridge waveguide.

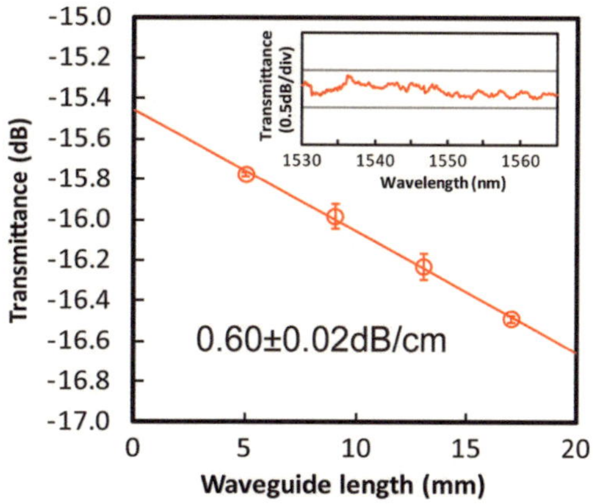

Figure 7. Measured transmittances of waveguides with 440-nm-thick a-Si:H cores. The error bars show the standard deviations (±σ) of the measured values. The inset shows the wavelength dependence of the waveguide in the C-band for a 17-mm-long shallow-ridge waveguide.

3. Vertical interlayer transition devices

3.1. VIT device structure and design

3.1.1. Structure of the VIT device

An optical VIT is required to construct 3D PICs. The optical connection of vertically stacked optical circuit layers dramatically enhances the architecture flexibility of a circuit. **Figure 8** shows our concept of a VIT device [25]. The lower Si waveguide is optically coupled with the upper a-Si:H waveguide through the silicon oxynitride (SiON) secondary core (SC). The upper and lower waveguides are isolated with a SiO_2 interlayer. The SiON SC is located on the SiO_2 interlayer, meaning that the lower Si waveguide is spatially separated from the SC. Here, a wide, thin SiON SC is used because it enables efficient optical coupling from the lower Si tapered waveguide to the overlying SC. The evanescent field of the SC expands outward from the SC core. With regard to the coupling of the SC with the upper a-Si:H tapered waveguide, the coupling mechanism is similar to that of a typical inverse-taper spot-size converter [26–29]. Such a VIT device is predicted to achieve efficient, broadband and polarization-insensitive vertical light coupling. We employed a knife-edge taper technique that can form unique inverse tapers using a CMOS backend-compatible process. The tapered waveguides have a trapezoidal cross-section. Near the taper end, the cross-section becomes triangular, with its height gradually decreasing toward the end. **Figure 9** shows a schematic diagram of the knife-edge tapered waveguide together with its optical mode fields calculated by the FDM [30]. The details of the fabrication process are described in Section 3.2.

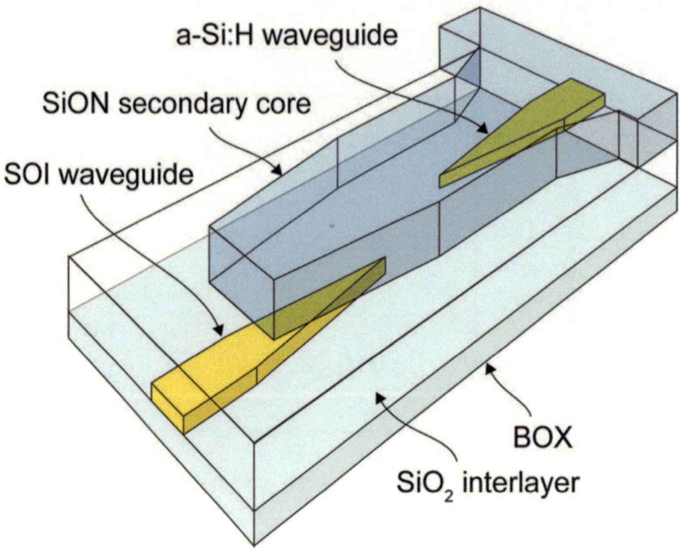

Figure 8. Schematic diagram of a VIT device.

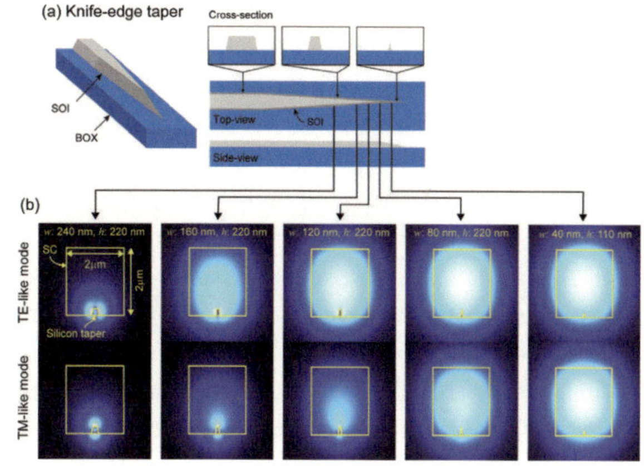

Figure 9. (a) Schematic diagrams of the knife-edge tapered waveguide. (b) Main electrical fields for the optical modes of the knife-edge tapered waveguide at 1.55 μm calculated using FDM. The refractive index of the secondary core was assumed to be 1.535, and the cladding material was assumed to be SiO_2.

3.1.2. Design of the VIT device

The dimensions of the SiON SC were optimized to minimize the insertion losses for both q-TE and q-TM modes of light at 1.55 μm using a numerical analysis based on an eigenmode expansion (EME) method. **Figure 10** presents the structural parameters used in the simulation. The Si waveguide was 400 nm wide and 220 nm high, and the distance between the two tapered tips of the waveguides was 50 μm. The upper and lower waveguides were isolated by a 0.6-μm-thick SiO_2 interlayer. The interlayer thickness satisfied design requirements for optical loss and crosstalk at orthogonal crossings between vertically adjacent Si waveguides [31]. The SC had a tapered structure near its end to lower loss caused by refractive index discontinuity. The tip width was 0.8 μm, which was a critical dimension for processing. The refractive indices of Si, SiON, and SiO_2 used in the simulation were 3.48, 1.62, and 1.44, respectively. In the actual devices, knife-edge tapers were used, but the simulation used standard inverse tapers, as depicted in **Figure 10**.

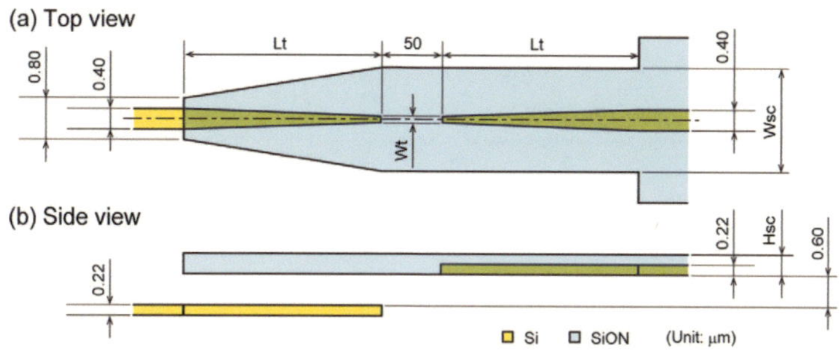

Figure 10. (a) Top view and (b) side view of the simulation model. The dimensions in these views are given in micrometers.

Figure 11(a) displays the transmittances calculated for various SiON SC heights (Hsc) when the SC width (Wsc) was fixed at 2.0 μm. The Si taper length (Lt) was set at 200 μm while the tip width of the Si tapers (Wt) was fixed at 0 μm; that is, ideal horizontally inverse tapers were assumed. Maximum transmittances for both the q-TE and TM modes were obtained at Hsc of around 0.4 μm. A thick SC (over 0.6 μm) decreases the transmittance because of smaller overlap of the optical field between the SC and lower Si taper. In contrast, **Figure 11(b)** shows the width of the SC is insensitive to the transmittance when Hsc is 0.4 μm, Lt is 200 μm, and Wt is 0 μm. According to the above EME simulations, the transition losses are minimized to 0.013 and 0.025 dB for the q-TE and TM modes, respectively, when Hsc is 0.4 μm and Wsc is 2.0 μm. The dependence of transmittance on Lt when Hsc is 0.4 μm, Wsc is 2.0 μm, and Wt is 0 μm is illustrated in **Figure 11(c)**. A taper that is more than 150 μm long is required to achieve low loss (<1%) and polarization-independent operation. In the above simulation, Wt was assumed to be 0 μm. However, the width of a taper tip is actually a finite value. **Figure 11(d)** summarizes the dependence of transmittance on Wt. A knife-edge tapered structure could achieve a tip width of less than 50 nm [28–30]. Assuming Wt is 50 nm, the worst transition loss caused by the finite tip of the knife-edge taper would be 0.054 and 0.18 dB for q-TE and TM modes, respectively.

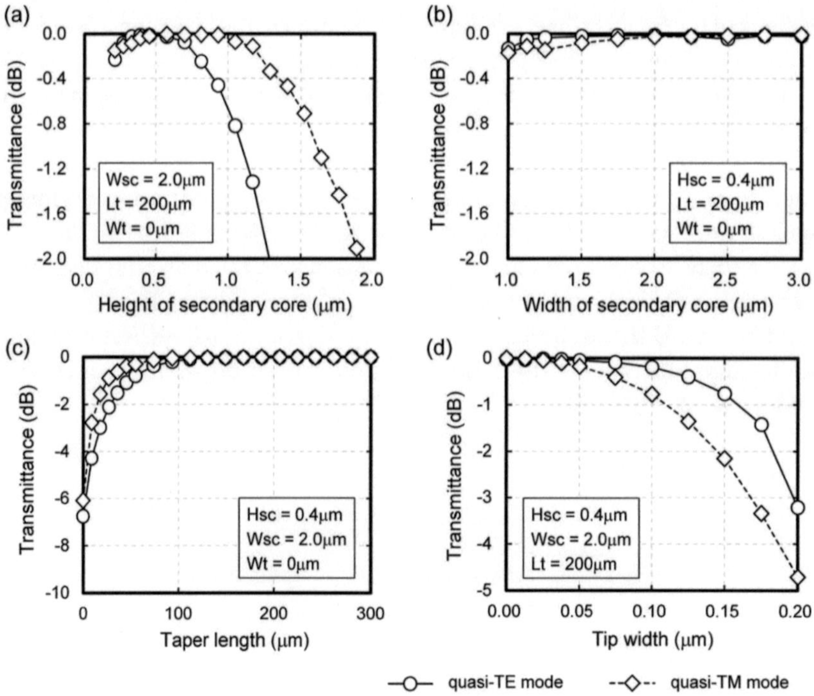

Figure 11. Calculated transmittances of VIT devices as a function of (a) SiON secondary core (SC) height (Hsc) when SC width (Wsc) = 2.0 μm, (b) Wsc when Hsc = 0.4 μm, (c) taper length (Lt) when Wsc = 2.0 and Hsc = 0.4 μm, and (d) tip width of the Si tapers (Wt) when Wsc = 2.0 and Hsc = 0.4 μm. Lt was fixed at 200 μm except in (c), while Wt was 0 μm except in (d). Circles on the solid lines are the transmittances for the q-TE mode, and diamonds on the dashed lines are the transmittances for the q-TM mode.

http://dx.doi.org/10.5772/63374

Figure 12 presents the simulated light propagation in a VIT device that incorporates the knife-edge tapered waveguides with the optimized structural parameters from above. The sidewall angle of the Si waveguide tapers was assumed to be 80°. The light in the lower Si waveguide was transferred to the upper Si waveguide with no apparent loss for both the q-TE and q-TM modes.

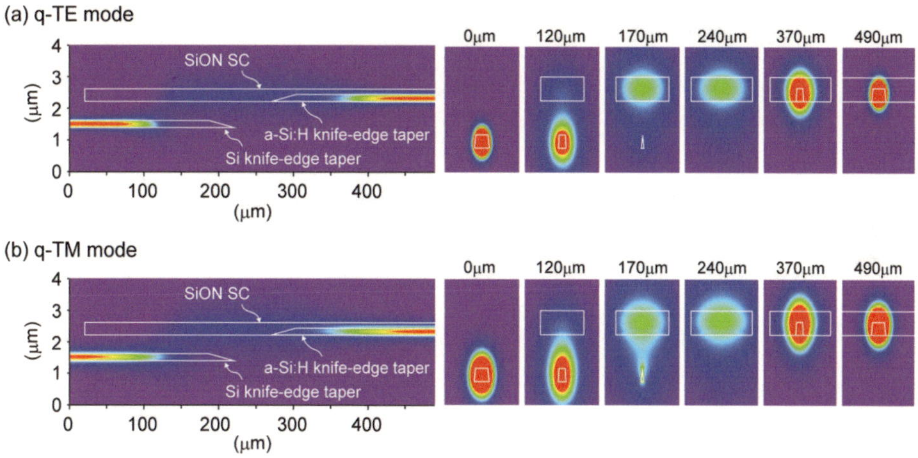

Figure 12. Side and cross-sectional views of the simulated light propagation in a VIT device with knife-edge tapered waveguides for (a) q-TE and (b) q-TM modes. The figure above each cross-sectional view shows its position shown in the side view.

3.2. Fabrication of a VIT device

First, the fabrication process for the knife-edge tapered waveguides is described. The knife-edge tapered waveguides were produced by a combination of a double-patterning technique and angled sidewall Si etching. **Figure 13** outlines the principle of the knife-edge tapering approach. In the initial wafer processing, a Si waveguide is formed, accompanied with an additional parallelogram structure. Subsequently, Si in the additional structure is partly removed under the same process conditions, resulting in the formation of a knife-edge tapered structure. In the middle of the taper, the edge of the second pattern is on the flat top of the Si parallelogram structure. In this case, the Si remaining after the etching has a trapezoidal cross-section. Near the taper tip, the edge of the second pattern is at the sloped sidewall. In this case, the remaining Si has a triangular cross-section, the height of which tapers down toward the end. Combining these two individual patterns enabled us to form ultrafine patterns beyond the photolithographic resolution limit. Furthermore, this technique possesses a large fabrication tolerance. Irrespective of the overlay misalignment, the second pattern crosses the first pattern because of the margin of the first pattern. The sidewalls of the etched Si were angled by controlling the etching chemistry; that is, the mixing ratio of SF_6 to C_4F_8. **Figure 14(a)** shows SEM images of the cross-sections of etched silicon with mixing ratios of SF_6 to C_4F_8 of 33 and 25%, which gave sidewall angles of ~80° and ~60°, respectively. The AFM image of the fabricated knife-edge taper in **Figure 14(b)** reveals that an ultra-narrow taper structure was successfully formed.

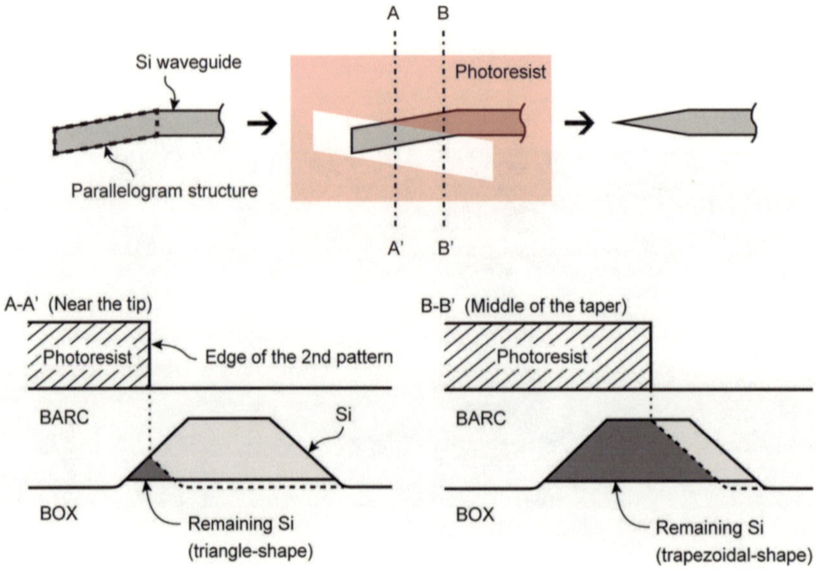

Figure 13. Schematic diagrams of the principle of knife-edge tapering. BARC is bottom anti-reflective coating.

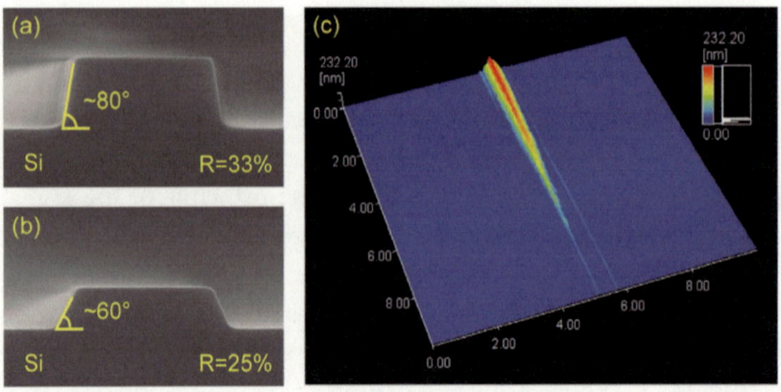

Figure 14. Cross-sectional SEM images of the etched silicon with mixing ratios of SF_6 to C_4F_8 of (a) 33% and (b) 25%. (c) AFM image of the fabricated knife-edge taper near its end.

VIT devices were fabricated on an SOI wafer with a 220-nm-thick top Si layer and a 2-μm-thick BOX layer. Si waveguides that were terminated with the knife-edge tapers were fabricated using an i-line stepper and ICP-RIE. This step was followed by SiO_2 deposition at 350°C by PECVD followed by top surface planarization by CMP. SiO_2 was again deposited by PECVD to adjust the SiO_2 thickness to precisely 0.6 μm. Subsequently, a 220-nm-thick a-Si:H film was deposited using our low-loss a-Si:H thin-film technology described in Section 2. The a-Si:H waveguides were fabricated by the same procedure that was used to produce the lower SOI layer. A 0.4-μm-thick SiON film was deposited at 350°C by PECVD. SiON SCs with a width of 2 μm were then fabricated using an i-line stepper and fluoroform-based dry etching. Refractive indices of the deposited a-Si:H and SiON measured using ellipsometry were 3.50 and 1.62, respectively. Finally, the wafer was covered with a 1.5-μm-thick SiO_2 film that served as a cladding layer.

3.3. Results and discussion

The fabricated VIT devices were evaluated in the C-band wavelength range. The insertion losses of the fabricated VIT devices were extracted from their transmittance differences between the optical paths of different numbers of VIT devices. That is, paths with 2, 6, 10, and 14 VIT devices were investigated. All paths used Si and a-Si:H waveguides of the same length. An ASE light source was used to measure the transmittance of the devices in the wavelength range from 1530 to 1565 nm. The ASE light was polarized and then the polarized light was sent using a polarization maintaining fiber. The light was fed into each device chip using condenser lenses. The output light was collected by condenser lenses and measured by an optical spectrum analyzer. **Figure 15** shows the insertion losses of the VIT devices together with the measured transmittances of the optical paths with different numbers of VIT devices. The measured insertion losses at 1550 nm were 0.87 ± 0.06 and 0.79 ± 0.06 dB for the q-TE and q-TM modes, respectively. These values correspond to transition efficiencies of 82 and 83% for the q-TE and q-TM modes, respectively. Also, the wavelength dependences of the insertion losses were less than ~0.5 dB in the C-band for both polarizations. In the shorter wavelength region, the insertion loss increased slightly. This tendency contrasts with the EME simulation results that predicted flat wavelength dependences in the C-band. The slight increase is caused by the absorption of the SiON film around 1510 nm. This absorption originates from the higher-order vibrational modes of the N-H bonds in the SiON film fabricated by low-temperature PECVD.

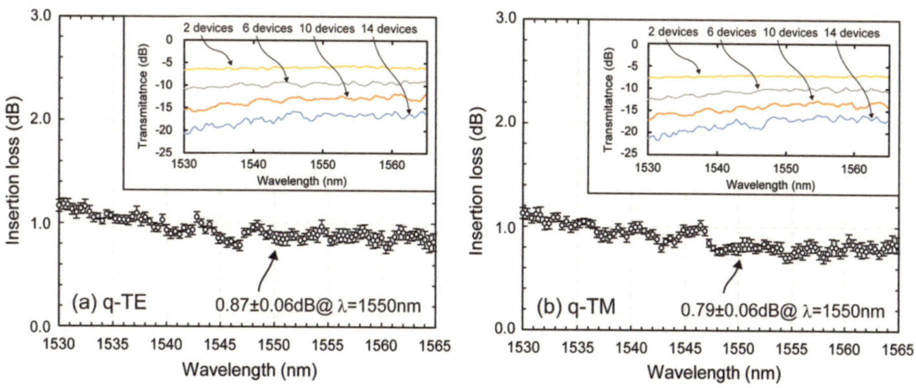

Figure 15. Measured insertion losses of VIT devices for (a) q-TE and (b) q-TM modes. The insets show the measured transmittances of the optical paths when they contain 2, 6, 10, and 14 VITs. Each transmittance value is averaged from the same four optical paths. The error bars show the standard errors.

The experimental losses are larger than those predicted by the EME simulation. To investigate these discrepancies, we performed numerical analyses with varying refractive indices of the a-Si:H and the SiON. The analyses revealed that the discrepancies cannot be explained by the refractive index variation of a-Si:H. Actually, our experiments indicated that the deposition of a-Si:H films is reasonably reproducible, so the variation of refractive index is typically less than ±2%. Even when the refractive index of a-Si:H fluctuates by ±5%, the numerical analyses suggested the loss deterioration of the VIT devices was less than 0.01 dB. Because the measured refractive index of a-Si:H was 3.50, the deviation from the value (3.48) used for the device design should be negligible. However, the numerical analyses showed that loss deterioration caused by variation of the refractive index of SiON would be considerable when its refractive index is less than 1.55, as shown in **Figure 16**. However, as long as the refractive index of SiON exceeds 1.55, low transition loss can be obtained for both polarizations. Because measured refractive index of SiON was 1.62, and is unlikely to drop below 1.55, the refractive index variation of SiON does not affect the performance of the VIT devices. Overall, the discrepancies between simulated and experimental losses of the VIT devices cannot be explained by the variations of the refractive indices of either a-Si:H or SiON. Furthermore, additional loss of the VIT devices caused by the finite tip width makes only a minor contribution (see **Figure 11(d)**). Therefore, we concluded that these discrepancies originate mainly from the waveguide losses of the Si and SiON waveguides. It should be noted that these numerical analyses indicate the VIT devices are more robust against unwanted variation of refractive index than interlayer coupling devices based on grating couplers, which are sensitive to the refractive indices of both waveguide and cladding materials.

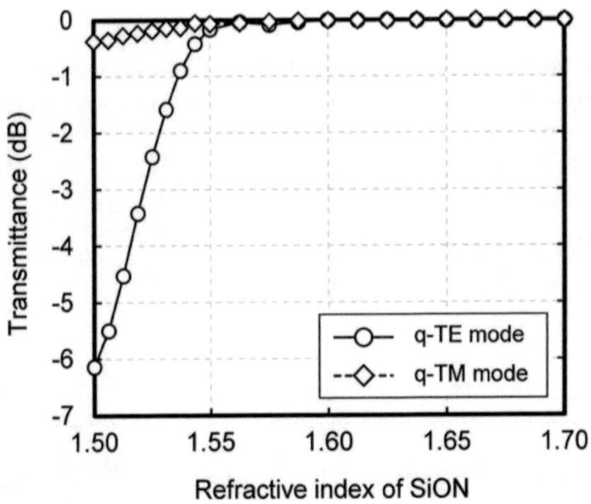

Figure 16. Calculated transmittances of the VIT devices as a function of the refractive index of SiON. The structural parameters are as follows: Hsc = 0.4 μm, Wsc = 2.0 μm, Lt = 200 μm, and Wt = 0 μm. The refractive index of a-Si:H was set at 3.48 to match that of the SOI.

Table 2 compares the performance of optical cross-coupling devices using a-Si:H deposited at low temperature that have been reported to date. All the devices are highly efficient. The vertical directional coupler has the most compact device, and its fabrication is simple, but the interlayer isolation is insufficient for 3D PICs [31]. The grating coupler effectively isolates the layers, but its bandwidth is relatively narrow and grating couplers have strong polarization dependence. In addition, metal mirrors are required to increase the efficiency of these couplers, resulting in complex fabrication. The VIT devices showed broadband, polarization-insensitive performance. However, long tapered waveguides are required in the VIT devices to suppress the losses and their fabrication process is complex because of the SC. Although the taper length was 200 μm in our work, the length of the tapers can be shortened to 150 μm with only a small increase of loss (see **Figure 11(c)**), resulting in a total device length of 350 μm. A parabolic taper structure would be advantageous to make the taper length shorter while keeping the mode conversion loss low [32].

Reference	[33]	[34]	Our work
Type	Inverse tapers	Grating coupler + mirrors	Inverse tapers + secondary core
Interlayer	200-nm-thick SiO2	1000-nm-thick SiO2	600-nm-thick SiO_2
Efficiency	~90% (TE)	83% (TE)	82% (TE), 83% (TM)
Polarization	Single	Single	Both
Length	30 μm	114 μm	450 μm
1-dB Bandwidth	Not described	~20 nm	> 35 nm
Process	Simple	Complex	Complex

Table 2. Performance comparison of optical cross-coupling devices with vertically stacked optical circuits.

In our work, VIT devices were demonstrated on SOI wafers. If the lower Si layer on the BOX was replaced with a-Si:H on SiO_2 deposited at low temperature, the VIT device performance could be equivalent to that of the VIT devices reported in this section. This means that these VIT devices can be realized on SiO_2-passivated CMOS chips. Therefore, 3D on-chip optical interconnects with efficient, broadband, and polarization-insensitive VIT functionality that are compatible with CMOS metal wiring layers are potentially available.

4. Conclusions

In this chapter, we described our research on a-Si:H photonics. Importantly, a-Si:H photonics is promising to achieve CMOS backend integration of optical interconnects because a-Si:H with ultra-low loss can be obtained at a deposition temperature of less than 400°C. Using our high-quality Si thin-film technology, we demonstrated a-Si:H waveguides with ultra-low loss and high-performance VIT devices. Also, electrically driven refractive index changes in a-Si:H were realized by introduction of μc-Si:H. These device technologies provide a fundamental platform for design of 3D PICs. Further development of a-Si:H photonics will drive the realization of on-chip optical interconnects in the future.

Author details

Ryohei Takei

Address all correspondence to: r.takei@aist.go.jp

National Institute of Advanced Industrial Science and Technology (AIST), Tsukuba, Japan

References

[1] R. A. Street. Technology and Application of Amorphous Silicon. Springer; 2000.

[2] K. Narayanan and S. F. Preble. Optical nonlinearities in hydrogenated-amorphous silicon waveguides. Opt. Express. 2010;18(9):8998–9005. DOI: 10.1364/OE.18.008998.

[3] Y. Shoji, T. Ogasawara, T. Kamei, Y. Sakakibara, S. Suda, K. Kintaka, H. Kawashima, M. Okano, T. Hasama, H. Ishikawa, and M. Mori. Ultrafast nonlinear effects in hydrogenated amorphous silicon wire waveguide. Opt. Express. 2010;18(6):5668–5673. DOI: 10.1364/OE.18.005668

[4] C. Lacava, P. Minzioni, E. Baldini, L. Tartara, J. M. Fedeli, and I. Cristiani. Nonlinear characterization of hydrogenated amorphous silicon waveguides and analysis of carrier dynamics. Appl. Phys. Lett. 2013;103(14).

[5] B. Kuyken, H. Ji, S. Clemmen, S. K. Selvaraja, H. Hu, M. Pu, M. Galili, P. Jeppesen, G. Morthier, S. Massar, L. K. Oxenløwe, G. Roelkens, and R. Baets. Nonlinear properties of and nonlinear processing in hydrogenated amorphous silicon waveguides. Opt. Express. 2011;19(26).

[6] K.-Y. Wang, K. G. Petrillo, M. A. Foster, and A. C. Foste. Ultralow-power all-optical processing of high-speed data signals in deposited silicon waveguides. Opt. Express. 2012;20(22).

[7] S. Suda, K. Tanizawa, Y. Sakakibara, T. Kamei, K. Nakanishi, E. Itoga, T. Ogasawara, R. Takei, H. Kawashima, S. Namiki, M. Mori, T. Hasama, and H. Ishikawa. Pattern-effect-free all-optical wavelength conversion using a hydrogenated amorphous silicon waveguide with ultra-fast carrier decay. Opt. Lett. 2012;37(8).

[8] K. Narayanan, A. W. Elshaari, and S. F. Preble. Broadband all-optical modulation in hydrogenated-amorphous silicon waveguides. Opt. Express. 2010;18(10).

[9] B. Kuyken, S. Clemmen, S. K. Selvaraja, W. Bogaerts, D. Van Thourhout, P. Emplit, S. Massar, G. Roelkens, and R. Baets. On-chip parametric amplification with 26.5 dB gain at telecommunication wavelengths using CMOS-compatible hydrogenated amorphous silicon waveguides. Opt. Lett. 2011;36(4).

[10] G. Cocorullo, F. G. Della Corte, R. De Rosa, I. Rendina, A. Rubino, and E. Terzini. Amorphous silicon-based guided-wave passive and active devices for silicon integrated optoelectronics. IEEE J. Sel. Top. Quantum Electron. 1998;4(6).

[11] A. Harke, M. Krause, and J. Mueller. Low-loss singlemode amorhpous silicon waveguides. Electron. Lett. 2005;41(25).

[12] S. K. Selvaraja, E. Sleeckx, M. Schaekers, W. Bogaerts, D. V. Thourhout, P. Dumon, and R. Baets. Low-loss amorphous silicon-on-insulator technology for photonic integrated circuitry. Opt. Commun. 2009;282(9).

[13] S. Zhu, G. Q. Lo, and D. L. Kwong. Low-loss amorphous silicon wire waveguide for integrated photonics: effect of fabrication process and the thermal stability. Opt. Express. 2010;18(24):25283–25291.

[14] R. Sun, J. Cheng, J. Michel, and L. Kimerling. Transparent amorphous silicon channel waveguides and high-Q resonators using a damascene process. Opt. Lett. 2009;34(15).

[15] J. Kang, Y. Atsumi, M. Oda, T. Amemiya, N. Nishiyama, and S. Arai. Low-loss amorphous silicon multilayer waveguides vertically stacked on silicon-on-insulator substrate. Jpn. J. Appl. Phys. 2011;50(12R).

[16] Z. E. Smith and S. Wagner. Band tails, entropy, and equilibrium defects in hydrogenated amorphous silicon. 1987;59(6):688–691.

[17] R. Takei, S. Manako, E. Omoda, Y. Sakakibara, M. Mori, and T. Kamei. Sub-1 dB/cm submicrometer-scale amorphous silicon waveguide for backend on-chip optical interconnect. Opt. Express. 2014;22(4):4779–4788.

[18] T. Kamei, N. Hata, A. Matsuda, T. Uchiyama, S. Amano, K. Tsukamoto, Y. Yoshioka, and T. Hirao. Deposition and extensive light soaking of highly pure hydrogenated amorphous silicon. Appl. Phys. Lett. 1996;68(17):2380.

[19] D. K. Sparacin, R. Sun, A. M. Agarwal, M. A. Beals, J. Michel, and L. C. Kimerling. Low loss amorphous silicon channel waveguides for integrated photonics. In: 3rd International Conference on Group IV Photonics; Ottawa, Canada. IEEE; 2006. p. 255–257.

[20] B. Han, R. Orobtchouk, T. Benyattou, P. R. A. Binetti, S. Jeannot, J. M. Fedeli, and X. J. M. Leijtens. Comparison of optical passive integrated devices based on three materials for optical clock distribution. In: ECIO 07; Copenhagen, Denmark. 2007. p. 1–4.

[21] S. Rao, F. G. D. Corte, and C. Sumonte. Low-loss amorphous silicon waveguides grown by PECVD on indium tin oxide. J. Europ. Opt. Soc. Rap. Public. 2010;5:10039s.

[22] R. Suzuki, Y. Kobayashi, T. Mikado, A. Matsuda, P. J. Mcelheny, S. Mashima, H. Ohgaki, M. Chiwaki, T. Yamazaki, and T. Tomimasu. Characterization of hydrogenated amorphous silicon films by a pulsed positron beam. Jpn. J. Appl. Phys. 1991;30(10): 2438–2441.

[23] K. Furuya, K. Nakanishi, R. Takei, E. Omoda, M. Suzuki, M. Okano, T. Kamei, M. Mori, and Y. Sakakibara. Nanometer-scale thickness control of amorphous silicon using

isotropic wet-etching and low loss wire waveguide fabrication with the etched material. Appl. Phys. Lett. 2012;100:251108.

[24] T. Kamei, P. Stradins, and A. Matsuda. Effects of embedded crystallites in amorphous silicon on light-induced defect creation. Appl. Phys. Lett. 1999;74:1707.

[25] R. Takei, Y. Maegami, E. Omoda, Y. Sakakibara, M. Mori, and T. Kamei. Low-loss and low wavelength dependence vertical interlayer transition for 3D silicon photonics. Opt. Express. 2015;23(14):18602–18610.

[26] T. Tsuchizawa, K. Yamada, H. Fukuda, T. Watanabe, S. Uchiyama, and S. Itabashi. Low-loss Si wire waveguides and their application to thermooptic switches. Jpn. J. Appl. Phys. 2006;45(8B):6658–6662.

[27] Y. Wakayama, T. Kita, and H. Yamada. Optical crossing and integration using hybrid si-wire/silica waveguides. Jpn. J. Appl. Phys. 2011;50(4S):04DG20.

[28] R. Takei, E. Omoda, M. Suzuki, S. Manako, T. Kamei, M. Mori, and Y. Sakakibara. Ultranarrow silicon inverse taper waveguide fabricated with double-patterning photolithography for low-loss spot-size converter. Appl. Phys. Express. 2012;5:052202.

[29] Y. Maegami, R. Takei, E. Omoda, T. Amano, M. Okano, M. Mori, T. Kamei, and Y. Sakakibara. Spot-size converter with a SiO_2 spacer layer between tapered Si and SiON waveguides for fiber-to-chip coupling. Opt. Express. 2015;23(16):21287–21295.

[30] R. Takei, M. Suzuki, E. Omoda, S. Manako, T. Kamei, M. Mori and Y. Sakakibara. Silicon knife-edge taper waveguide for ultralow-loss spot-size converter fabricated by photolithography. Appl. Phys. Lett. 2013;102:101108.

[31] K. Furuya, R. Takei, T. Kamei, Y. Sakakibara, and M. Mori. Basic study of coupling on three-dimensional crossing of Si photonic wire waveguide for optical interconnection on inter or inner chip. Jpn. J. Appl. Phys. 2012;51(4S):04DG12.

[32] V. R. Almeida, R. R. Panepucci, and M. Lipson. Nanotaper for compact mode conversion. Opt. Lett. 2003;28(15):1302–1304.

[33] R. Sun, M. Beals, A. Pomerene, J. Cheng, C.-Y. Hong, L. Kimerling, and J. Michel. Impedance matching vertical optical waveguide couplers for dense high index contrast circuits. Opt. Express. 2008;16(16):11682–11690.

[34] J. H. Kang, Y. Atsumi, Y. Hayashi, J. Suzuki, Y. Kuno, T. Amemiya, N. Nishiyama, and S. Arai. 50 Gbps data transmission through amorphous silicon interlayer grating couplers with metal mirrors. Appl. Phys. Express. 2014;7(3):032202.

Chapter 7

Amorphous Hydrogenated Carbon Films with Diamond-Like and Polymer-Like Properties

Elena A Konshina

Additional information is available at the end of the chapter

http://dx.doi.org/10.5772/62704

Abstract

Results of the study of structural features and optical properties of thin films of amorphous hydrogenated carbon (a-C:H) films prepared by plasma-activated chemical vapor deposition of various hydrocarbon precursors are reviewed. The effect of different factors on the rate of a-C:H films deposition in a DC glow discharge with the magnetron plasma localized near the anode such as voltage, discharge power, gas pressure, relative content of an inert gas in the mixture with a hydrocarbon and other is analyzed. It is shown that the refractive index of a-C:H films can be changed in the interval 2.35–1.55 by increasing the deposition rate and the choice of the appropriate hydrocarbon precursor. The features of the vibration spectra of the diamond-like and polymer-like films are discussed. The correlations of the structural peculiarities and of the optical absorption edge, gap width, and conductivity as well as the absorption spectra in visible region and the ratio of the fundamental bands in Raman scattering spectra are estimated. Examples of using the optical properties of the a-C:H films are given.

Keywords: a-C:H films, deposition, refractive indexes, vibration spectra, optical gap

1. Introduction

Interest in the study of thin films of amorphous hydrogenated carbon (a-C:H) obtained by a plasma-activated chemical vapor deposition (CVD) is maintained for several decades. The reason for this is the possibility of variations in the properties of the films in a broad interval from diamond-like carbon up to polymer-like films. This opens up great opportunities for their practical application. The glow discharge plasma is produced with RF-biased in [1–6] or DC in [7–9] diode-type systems. The efficiency of gas ionization can be improved by placing a grid negatively biased with respect to the RF plasma near the substrate in [10], by using a magnetic field perpendicular to the RF plasma electric field in [11], and also by using a dual microwave electron cyclotron resonance RF plasma by applying an independently controlled RF substrate bias voltage in [12].

In this chapter, the results of the study of the structural features and optical properties of thin films prepared by CVD process in a DC glow discharge with the magnetron plasma localized near the anode are discussed. The effect of different factors on a rate of a-C:H films deposition such as voltage, discharge power, gas pressure, relative content of an inert gas in the mixture with a hydrocarbon and other is analyzed. Ellipsometry method was used for a comparative analysis of optical constants of a-C:H films prepared from different precursors. Features of the vibration spectra of a-C:H films according to their refractive index are discussed. The correlation of the absorption bands in the visible region of the spectra with of the optical gap and the Raman scattering spectra of the a-C:H films are explained. Examples of the use of diamond-like and polymer-like properties of the a-C:H films are shown.

2. Deposition of films in a DC glow discharge

2.1. Experimental setup for plasma-activated CVD

The a-C:H films are prepared by the method of CVD of hydrocarbons in the DC glow discharge using magnetron plasma localized at the anode in a quasi-closed volume [13]. The schematic representation of the device for the deposition of a-C:H films is shown in **Figure 1**.

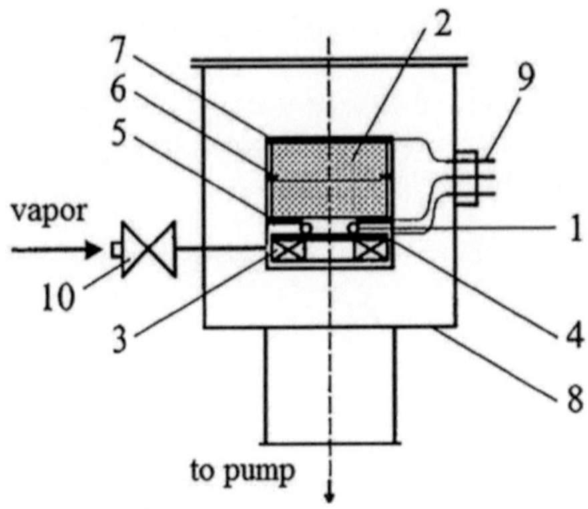

Figure 1. Schematic of the CVD device on DC with magnetron plasma: 1—localized plasma; 2—glow discharge plasma; 3—permanent magnet; 4—cathode; 5—anode; 6—additional electrode; 7—substrate holder; 8—vacuum chamber; 9—current leads; and 10—valve.

Localized plasma (1), produced by crossed magnetic and electric fields, sustains a glow discharge in volume (2). We used annular permanent magnet (3) and two planar annular electrodes: anode (4) and cathode (5). The magnetic field intensity near the cathode surface was about 600 G. Additional electrode (6) and substrate-carrying electrode (7) were mounted

http://dx.doi.org/10.5772/62704

on glass cylinders with a diameter of about 130 mm. The cylinder walls confine the plasma inside the quasi-closed volume of the vacuum chamber. The ring-shaped gap between the magnet and the cathode is used as a gas inlet. Such a design of the device provides uniform gas distribution and efficient gas consumption, as well as reduces the film contamination by foreign impurities. Cathode (4) and anode (5) are placed in the neighborhood of the permanent magnet to provide maximum intensity and uniformity of the magnetic field in the crossed field region. The spacing between cathode 4 and substrate holder (7) is about 50 mm in the absence of the additional electrode. This electrode is placed 30 mm away from the cathode.

The cathode was under the ground potential. The substrate holder was either negatively biased with a DC power unit or under the ground potential. The additional electrode was used as a substrate holder when a-C:H films were deposited at a glancing angle in [14]. The vacuum chamber was evacuated to $1–5 \times 10^{-3}$ Pa by a rotary backing pump and a turbo-molecular pump. When the voltage is applied to the electrodes, the high-density toroidal plasma arises as a result of the effective electron capture by a magnetic trap, followed by gas ionization in the discharge gap between the anode and the substrate holder. Ion collisions with the solid surface confining the plasma generate extra electrons which also take part in gas ionization. a-C:H films were deposited on polished copper and glass substrates covered by thin conducting and semiconducting layers. This device allows the deposition of films with a thickness nonuniformity of not more than 20% on substrates of diameter up to 100 mm. Dependences of the ion current *I* through substrate holder vs. acetylene pressure *P* in the vacuum chamber in **Figure 2** are shown the effect of the localized plasma generated in the region of the crossed electric and magnetic fields.

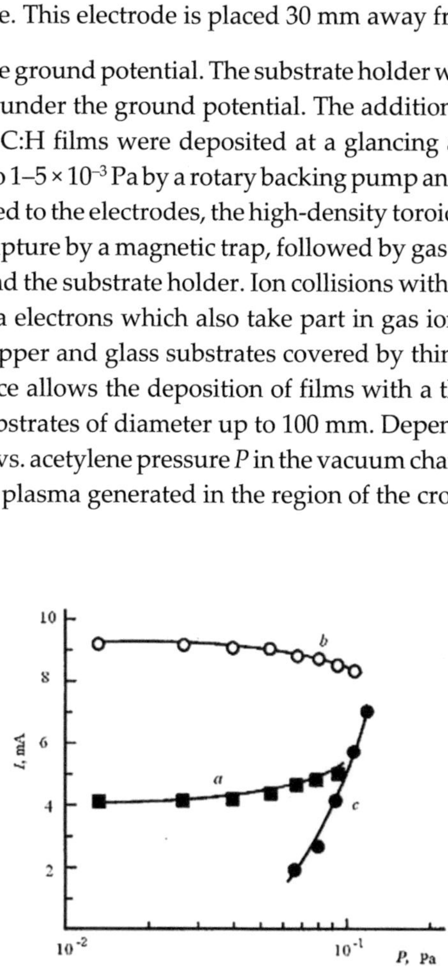

Figure 2. Dependencies of an ion current on the substrate holder vs. a pressure in the vacuum chamber. Glow discharge is sustained by the localized plasma (*a*, *b*) and without it (*c*). *U* = 600 (*a*), 800 (*b*), and 1200 V (*c*).

When the plasma is sustained by the plasma localized in the device, the ion current varies only slightly with the gas pressure (varying from 0.01 to 0.04 Pa) at a constant voltage *U* (curve *a* and curve *b* in **Figure 2**). An increase in the voltage from 600 to 800 V leads to a twofold increase in the current through the substrate holder. In the absence of the permanent magnet, the glow discharge was initiated at a higher pressure, 0.06 Pa (curve *c* in **Figure 2**), and at a higher voltage, 1200 V. In this case, the ion current depends more strongly on the gas pressure and increases with it. In comparison with standard diode-type systems, this device has a wider pressure range from 0.004 to 0.1 Pa and greatly extends the range of operating conditions for the deposition of a-C:H films in glow discharge plasma. The discharge power dissipated by positive ions at the substrate holder can vary between 1.2 and 44 W.

2.2. Factors governing the rate of a-C:H film deposition

Chemical reactions and physical processes at the surface of a-C:H films during their deposition in a low-temperature hydrocarbon plasma were considered in terms of the adsorbed layer model in [15]. This model assumes that CH_3 plasma radicals are physically adsorbed on the surface and then pass into the chemisorbed state as a result of cross-linking due to energetic ions. The surface coverage depends on the number of surface states and surface temperature. When a-C:H films are deposited with the described device, their surface is continuously bombarded by positive ions whose energy depends largely on the voltage. The ion energy may be high enough to consolidate the condensate either by cross-linking or by decomposing weakly bound particles with subsequent surface diffusion and desorption of the decomposition products. The balance of the processes conducive to film deposition and etching defines the deposition rate. The film deposition rate can also be varied by properly selecting the glow discharge parameters (gas pressure, ion current to the substrate, and voltage), which control the number of ions and their energy.

The deposition rate was defined as the ratio of the a-C:H film thickness to the deposition time. The thickness was measured by an MII-4M (Russia) micro-interferometer with an accuracy of 10%. The thickness of a-C:H films was found to be 0.1–0.3 μm. As the voltage increases from 700 to 1300 V, the deposition rate varies from 0.25 to 2 Å/s. The rate r of film deposition on the surface of the glass substrates in the acetylene plasma as a function of the voltage is shown in **Figure 3**.

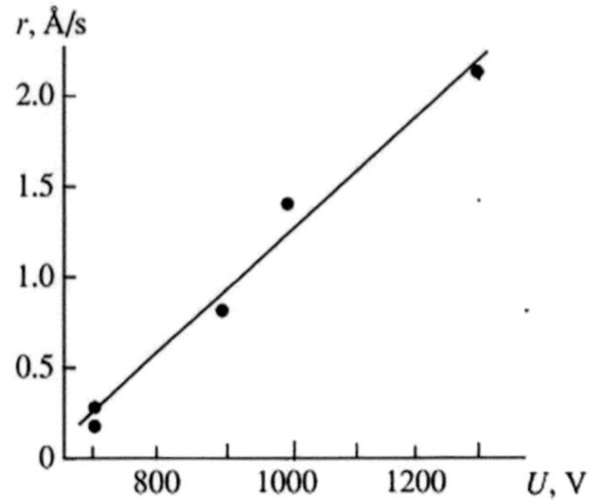

Figure 3. Rate of a-C:H films deposition on the glass substrates vs. the voltage.

Figure 4 illustrates the effect of surface conductivity on the deposition rate at an acetylene pressure in the vacuum chamber of about 0.05 Pa. For the same discharge power, the rate of film deposition on the surface of a transparent conducting indium–tin oxide (ITO) layer is one order of magnitude higher than that on the surface of a-Si:C:H semiconductor layer with a resistivity of about 10^{12} Ω cm.

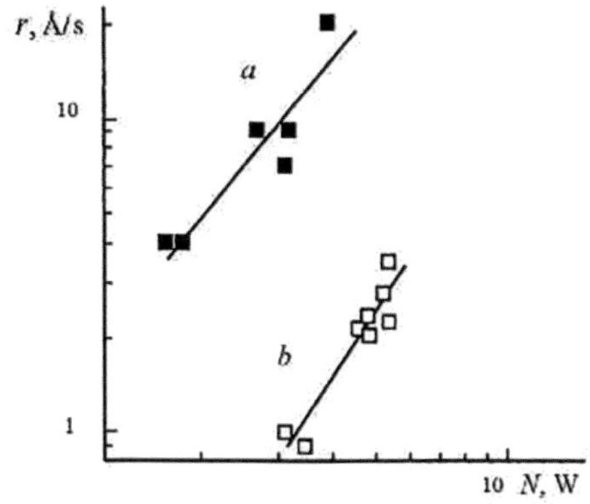

Figure 4. Rate of a-C:H film deposition on the ITO transparent conducting electrode (*a*) and a-Si:C:H semiconducting layer (*b*) at the acetylene pressure P = 0.05 Pa vs. discharge power N.

Space charge is produced on the substrate surface when it is bombarded by positive ions in a DC glow discharge plasma. Charge leakage from the surface depends on the substrate conductivity, as well as on the thickness and resistivity of the growing film. As the resistivity grows, the critical thickness of the film and the deposition rate decrease. Therefore, the surface conductivities of the substrate and condensate in glow discharge plasma should be considered

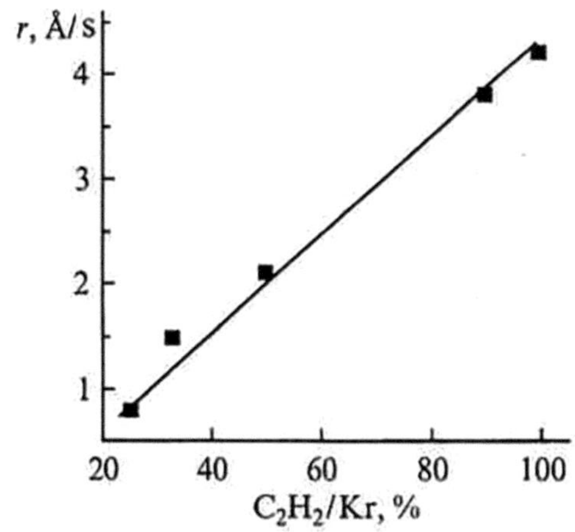

Figure 5. Rate of a-C:H film deposition on the copper substrates vs. acetylene volume content in the acetylene–argon mixture.

as important factors affecting the kinetics of a-C:H film deposition. The dependence of the deposition rate on the acetylene volume concentration in a mixture with krypton is shown in **Figure 5** for a constant discharge power of 1.8 W and a pressure of 0.05 Pa.

The addition of an inert gas to acetylene decreases the deposition rate of the a-C:H films on the copper substrates for the same pressure in the vacuum chamber from 4 to 0.5 Å/s. A film bombardment by inert gas ions etches the surface during the condensation and lowers the deposition rate.

3. Optical constants of a-C:H films

The method of ellipsometry enables to investigate the optical constants of thin films in the visible region. Ellipsometric studies have shown that the a-C:H films can have a high refractive index and transparent in the UV–visible range in [16, 17]. The dependences of the optical constants of a-C:H films on the parameters of the condensation process have been established by ellipsometry in [18–21]. The optical constants for the wavelength of 632.8 nm of the a-C:H films obtained by the method described above were determined using LEF-3M ellipsometer (Russia) [21]. The experimental setup consisted from polarizer–compensator–sample analyzer. A dual-zone null method at angles of incidence f of 50°, 60°, and 70° was used to calculate the ellipsometric parameters. To calculate a refractive index n, an extinction coefficients k, and a thickness d of the films on glass substrates with $n_s = 1.515$ and $k_s = 0$, the homogeneous isotropic layer model was used. The refractive indices and extinction coefficients as a function of the deposition rate of a-C:H films obtained from acetylene, toluene, and octane, as precursors, are shown in **Figures 6** and 7. Increasing the rate of film deposition in plasma from 0.2 to 2.5 Å/s was achieved by increasing the voltage in the range of 700–1300 V.

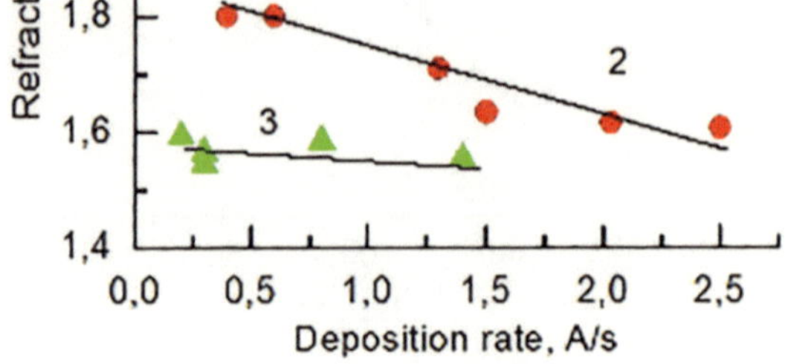

Figure 6. Refractive index of the a-C:H films obtained from (1) acetylene, (2) toluene, and (3) octane vs. the deposition rate.

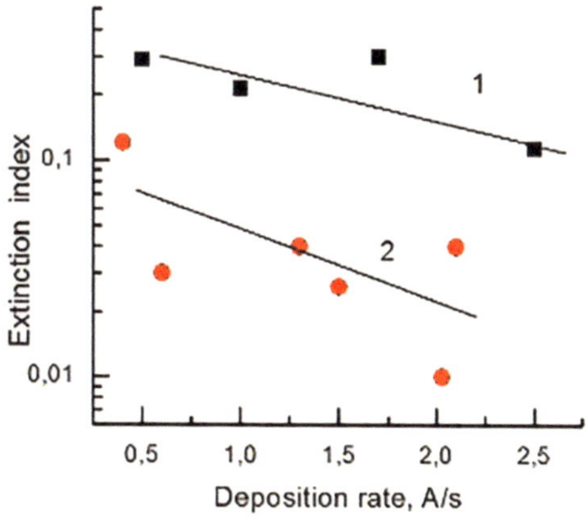

Figure 7. Extinction coefficients of the a-C:H films obtained from (1) acetylene and (2) toluene vs. the deposition rate.

The refractive index of a-C:H films obtained from acetylene and toluene decreases monotonically with increasing the deposition rate in **Figure 6**. Decreasing the extinction coefficients for these films can be seen in **Figure 7**. The films obtained from acetylene have the highest refractive indexes n = 2.35 and the highest extinction indexes k = 0.3 under the deposition rate equal to 0.5 Å/s. For films prepared from toluene, these indexes were equal to n = 1.8 and k = 0.1 in the same deposition rate. The films obtained from octane under the same conditions have the lowest values of n = 1.55–1.6 and k < 0.01 at a wavelength of 632.8 nm, and no dependence on the deposition rate is observed in **Figure 6**. These points to the dependence of the optical constants of the films from hydrocarbon precursors used for their deposition in the glow discharge plasma. It can be concluded that the optical constants of a-C:H films can be varied by changing the deposition rate in the plasma, as well as by selecting the precursor.

4. Vibration spectra of the a-C:H films

Infrared spectroscopy is widely used to study the optical absorption and structural features of a-C:H films in [22–29]. However, this method is not sufficiently sensitive for systems composed of a-C:H films and semiconductors with high refractive indices (Si and Ge) due to the significant interference effect. Comparative analysis of the IR spectra of multiple attenuated total internal reflection (MATIR) in the range 4000–1000 cm^{-1} was performed in [30]. The use of the MATIR method excludes the influence of the interference effect and makes it possible to record the vibration spectra of thin a-C:H films. A single-crystal germanium prism, which provided 12 reflections of IR radiation from a plane surface at an angle of 45°, served as the MATIR element. **Figure 8** shows the IR spectra of a-C:H films with the refractive index $n = 2.3$–2.4. These films were obtained from toluene (1), octane (2), and cyclohexane (3) vapors at pressures of 0.01–0.03 Pa and discharge powers of 10–13 W. It should be noted that the spectra of diamond-like C:H films obtained from various hydrocarbons have similar IR spectra in **Figure 8**. **Figure 9** shows the IR spectra of a-C:H films with the refractive indices $n = 1.64$ and

1.55, which were prepared at a vacuum-chamber pressure of ~0.3 Pa and a discharge power of 2 W from toluene (1) and octane (2) vapors, respectively.

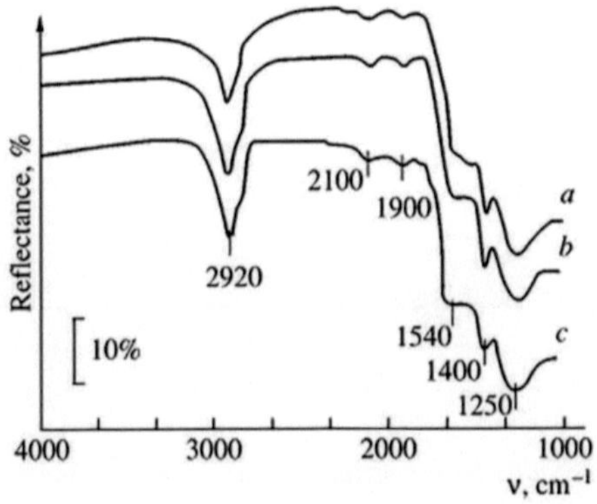

Figure 8. IR spectra of the diamond-like a-C:H films prepared from vapors of (1) toluene, (2) octane, and (3) cyclohexane.

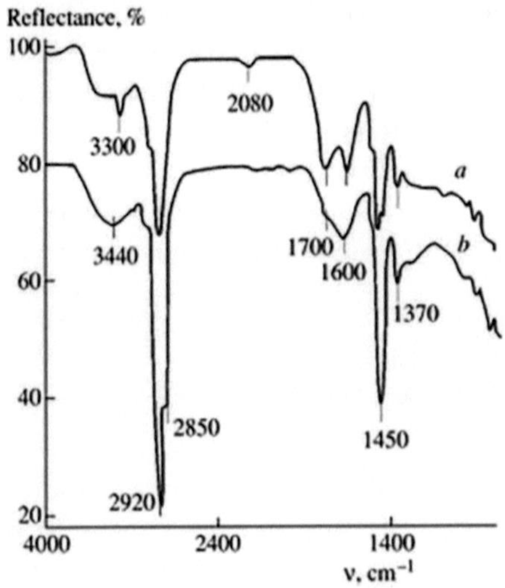

Figure 9. IR spectra of polymer-like a-C:H films prepared from vapors of (1) toluene and (2) octane.

As can be seen from **Figures 8** and **9**, an increase in the power by a factor of 5 (from 2 to 10 W) leads to significant changes in the IR spectra of the a-C:H films. The higher the discharge power, the higher the energy of the positive ions and their effect on the deposition process increase. The four functional branching structures are formed during the deposition of diamond-like films by decomposition of hydrocarbons in glow discharge plasma in the interaction of ions with the surface of the growing film. There are the stretching vibrations of C–H groups, carbonyl groups, single (C–C) and double (C=C) bonds, and bending vibrations of C–H groups in the spectra in **Figures 8** and **9**. In addition, weak peaks at 2100 and 1900 cm^{-1} (**Figure 8**) and 2080 cm^{-1} (**Figure 9**) due to the stretching vibrations of the C≡C bonds are observed in the spectra. The diamond-like films in **Figure 8** and polymer-like films in **Figure 9** have specific features in the IR absorption spectra. It should be noted that the IR spectra of the diamond-like a-C:H films (**Figure 8**) prepared from different hydrocarbons (octane, toluene, and cyclohexane) do not demonstrate any significant differences. They are similar to the spectra of the a-C:H films prepared from acetylene in [30]. The IR spectra of the polymer-like a-C:H films prepared from toluene and octane (**Figure 9**) are similar to previously measured spectra of polymer-like films prepared from benzene, toluene in [28], and acetylene in [29, 30]. The symmetric (ν^s) and asymmetric (ν^{as}) stretching vibrations and the symmetric (δ^s) and asymmetric (δ^{as}) bending vibrations of aromatic and olefin compounds, which observed in the IR spectra of the diamond- and polymer-like a-C:H films, are shown in **Table 1**.

Frequency, cm^{-1}	**Nature of vibration**	**Intensity of band**	
		Diamond-like films	**Polymer-like films**
~3400	ν (–OH)		medium
3300	ν (≡CH)		low
3100	ν (=CH) aromatic sp^2	low	low
3050	ν^{as} (=CH) aromatic sp^2	low	low
3000	ν(=CH) olefin sp^2	low	low
2920	ν^{as} (–CH) sp^3	high	high
2850	ν^s ($-CH_2$) sp^3	medium	medium
1900–2100	ν (C≡C)	very low	very low
~1700	ν (C=O)		high
1600	ν (C–C) aromatic		medium
1540	ν^{as} (C=C) polyene	medium	
1450	δ^{as} ($-CH_3$)		medium
1440	δ (–CH)	medium	
1370–1375	δ^s (CH_3)		medium
1250	ν (C–C) complex branching	medium	

Table 1. The frequencies, intensities, and nature of the vibrations in the IR spectra of the diamond- and polymer-like a-C:H films in the range 4000–1000 cm^{-1}.

A feature of the IR spectra of these films is the presence of bands at ~1250 cm^{-1} due to the stretching vibrations of C–C bonds at the branch points of four functional structures (**Table 1**). This band weakly manifests itself or is almost absent in the spectra of the polymer-like a-C:H films. The spectra of these films show strong absorption bands due to the stretching (~2920 cm^{-1}) and bending (~1450 cm^{-1}) vibrations of CH groups, as well as of carbonyl (~1700

cm^{-1}) and hydroxyl (~3400 cm^{-1}) groups in [27–30]. The reason of occurrence of the stretching vibrations of carbonyl and hydroxyl groups in the IR spectra is in the chemisorptions of water and oxygen from the environment into micropores of a-C:H films. The porosity of a-C:H films with a low refractive index can be up to 7% as the ellipsometric investigation of the films was shown in [21].

The content of bound hydrogen in polymer films is much more than the diamond-like films. The intensity of the absorption band in the range 3400–2600 cm^{-1} in the spectra of the polymer-like a-C:H films (**Figure 9**) is much higher than the spectra of the diamond-like films (**Figure 8**). The correlation between the refractive index of the a-C:H films and the integrated intensity of the band peaked at 2900 cm^{-1} has been shown in [27]. With a change in the refractive index from 1.55 to 2.4, the integrated intensity of the band attributed to the stretching vibrations of the CH groups exponentially decreases by an order of magnitude. Hence, a decrease in the content of bound hydrogen is accompanied by an increase in the refractive index of the a-C:H films, which indirectly proves that the film structure becomes denser.

5. Structural features of the a-C:H films

The known forms of amorphous carbon, including the various modifications of a-C:H, consist of carbon atoms in the sp^3-state which forms tetrahedral σ-bonds with four neighboring atoms, atoms in the sp^2-state which forms σ-bonds with three neighboring atoms in a plane, a weak π-bond perpendicular to this plane, and, to a lesser extent, atoms in the sp^1-state. According to the cluster model for the structure of amorphous carbon developed in [31–34], aromatic graphite π-bound atomic clusters consisting of sp^2-states are distributed within a sp^3-bound matrix. The π-states form a filled valence band, while the π^*-states correspond to an empty conduction band in the optical gap near the Fermi level. In accordance with the cluster model, the width of the optical gap depends on the sizes of the π-clusters and decreases as the fraction of sp^2-states which form them rises in [34] while the sp^3-bound matrix determines the tunnel barrier between them in [31]. It has been shown in [35, 36] that there is no difference in the nature of the states below and above the edge of the optical absorption edge and that the density of the π- and π^*-states close to the Fermi 2 eV.

Increasing the hydrogen concentration in an amorphous carbon structure leads simultaneously to a reduction in its equilibrium density and to a substantial change in the character of the clustering. This has been shown by studies of the stability of a-C:H systems and of their atomic and electronic structure as functions of the mass density and concentration of hydrogen using the molecular-dynamic density method in [37, 38]. The extent of the clusters is reduced by the introduction of sp^2-segments inside the stressed sp^3-matrix. The distribution of the clusters and of the segments which bind them affects the energy gap of a-C:H in [38]. The reduction of the gap is most likely a consequence of the splitting of the smallest π-bonded clusters and of the introduction of local π-electron systems into a stressed-bonded rigid lattice in which mixed bonds predominate.

5.1. Relationship of the Tauc parameters to the absorption spectra of a-C:H films

The optical gap in the various modifications of a-C:H depends on the conditions, under which it was produced in glow discharge plasma. The gap is made smaller when the energy delivered to the condensate is increased by raising the temperature of the carrier gas in [39] or substrate in [4, 20], by raising the power of the RF discharge [26], and by raising the voltage in [21, 40–42]. The conductivity of the a-C:H films increased when the optical gap decreased in [40].

For studying absorption spectra of a-C:H in the visible region, thin films were deposited from pure acetylene (C_2H_2) or a mixture of it with argon (Ar) at the ambient temperature. **Table 2** shows the conditions of preparation of the samples of the films. Samples *a* and *b* were prepared from pure acetylene with $P = 0.1$ Pa and $U = 600$ and 1200 V, respectively. Samples *c* and *d* were obtained with $P = 0.01$ Pa and $U = 900$ V from mixtures of acetylene with 50% and 70% argon.

Samples	*P*, Pa	Precursor	*U*, V	I_{600}/I_{800}	E_{max}, eV	E_T, eV	$B \times 10^4$ eV^{-1} cm^{-1}
a	0.1	C_2H_2	600	2.7	2.07	2.09	2.3
b	0.1	C_2H_2	1200	0.33	1.65	1.65	5.0
c	0.01	C_2H_2/50%Ar	900	0.4	1.65	1.09	7.9
d	0.01	C_2H_2/70%Ar	900	4.5	2.07	1.14	5.6

Table 2. The deposition conditions and parameters of the absorption spectra E_{max} and I_{600}/I_{800}, along with the Tauc parameters E_T and *B* for a-C:H films.

The spectral variations in the absorption α (λ) at wavelengths in the range 400–2400 nm were obtained from the reflection spectra of films with thicknesses of 0.3–0.5 μm deposited on polished copper substrates. These spectra can be resolved into a series of Gaussian-type bands. In the wavelength interval from 400 to 1000 nm, two absorption bands were identified with peaks at 600 ± 5 and 800 ± 40 nm in [43]. The ratio of the integrated intensities of these absorption bands (I_{600}/I_{800}) and the energies E_{max} of the peaks in the spectra are listed in **Table 2**.

The absorption edge of this a-C:H films, like that of other amorphous semiconductors, was adequately fit by the Tauc equation in [44]. The Tauc optical gap E_T was determined by extrapolating a plot of $(\alpha E)^{1/2}$ as a function of energy *E* as described in [42]. The reflection and transmission spectra of the films were recorded from 200 to 1000 nm on a Hitachi spectrophotometer. In these measurements, the a-C:H films with a thickness of 0.2 μm are deposited on quartz substrates. The natural absorption coefficient α of the films was calculated using the Lambert–Beer law. The resulting values of E_T and the slope *B* of the straight line drawn as a function of *E* for the a-C:H films are listed in **Table 2**. The a-C:H film obtained from pure acetylene with a minimal voltage of $U = 600$ V and a relatively high pressure ~0.1 Pa had the narrow band in the spectra with a peak at 600 nm and the ratio of $I_{600}/I_{800} = 2.7$ (sample *a* in **Table 2**). It is characterized by a wide optical gap of about 2.09 eV. Increasing the voltage glow discharge doubled ($U = 1200$ V) led to a change in the absorption spectra and the appearance of additional isolated peaks. The maximum was at 750 nm, and the ratio $I_{600}/I_{800} = 0.33$ (sample *b* in **Table 2**). Raising the discharge voltage also changes the Tauc parameters of the a-C:H films. E_T fell to 1.65 eV, while *B* increased. This indicates a rise in the density of states at the edge of the absorption band of the film. Dilution of the C_2H_2 with argon Ar leads to a significant change in the spectra of the a-C:H films.

The impact of the ions of the inert gas on the growing film (sample *c*) is equivalent to increasing a voltage of a glow discharge in the case sample *b* in **Table 2**. However, the magnitude of E_T decreased and the numerical value *B* increased compared with sample *b* of **Table 2**. The ratio of intensities of the absorption bands in the spectrum increased in favor of the band at 600 nm considerably as the result of diluting C_2H_2 with argon and decreasing the rate of the a-C:H film deposition up to 1 Å/s (a sample *d* in **Table 2**) in [43]. The width of the Tauc gap in the studied a-C:H films changes from ~2 to 1 eV. The constant *B* increases from 2.3×10^4 to 9.2×10^4 eV^{-1} cm^{-1} (**Table 2**), which indicates a rise in the density of states at the absorption edge of a-C:H with decreasing E_T. The nature of the π–π * transitions, which determine the optical gap width, is not clear from the plot of $(\alpha E)^{1/2}$ as a function of *E*. A comparison of the magnitudes of E_{max}, which corresponds to the energy of the most probable π–π * transition, and E_T showed that the a-C:H films with several peaks in their spectra have $E_{max} \approx E_T$ (**Table 2**, *a* and *b*), while the a-C:H films with a broad absorption spectrum have $E_{max} > E_T$ (samples *c* and *d* of **Table 2**). Varying the optical gap correlated with changing in a resistivity (ϱ) films. **Figure 8** shows the experimental data of ϱ and E_T versus the deposition rate of the a-C:H films. The values of ϱ and *r* are plotted in a logarithmic scale. Representative Tauc gap data changed from ~0.8 up to ~1.7 eV, and an exponential rise of ϱ of five orders of magnitude from ~10^8 to ~10^{13} Ω cm was observed at the same time in [45] (**Figure 10**).

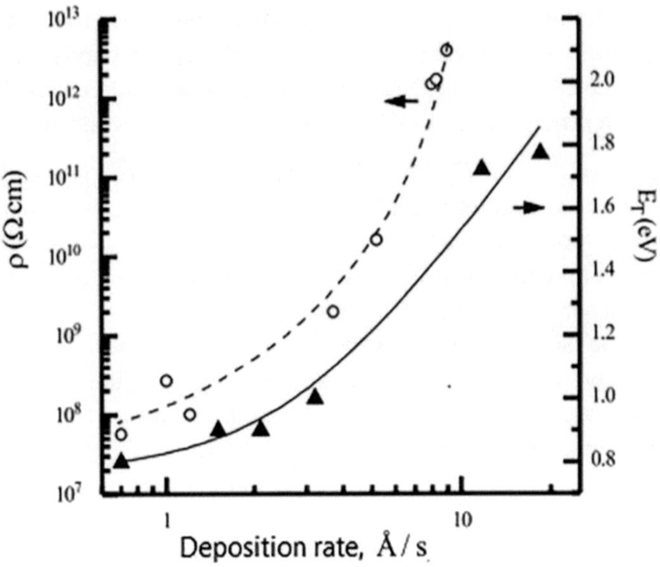

Figure 10. The electrical resistivity ϱ and Tauc gap E_T vs. the deposition rate of the a-C:H films.

The atomic structure of the diamond-like can be described as a stressed-bond rigid lattice, in which mixed bonds predominate with local electronic systems of π-bound atoms imbedded in it. The presence of CH-groups in the sp^3-state in the a-C:H structure reduces the extent of the π-bond clusters in [38], favors localization of their π-electrons, and raises the tunnel barrier between them in [32]. The distribution of the π-bond clusters in the sp^3-bound matrix is governed the Tauc parameters of the a-C:H films. The schemes of the density of states as a function of energy E generalizing the obtained data are shown in **Figure 11.** The structure of the optical gap of the a-C:H modification with $E_T > 1.5$ eV and a resistivity exceeding 10^{10} Ω cm can be illustrated by scheme *a* (**Figure 11**).

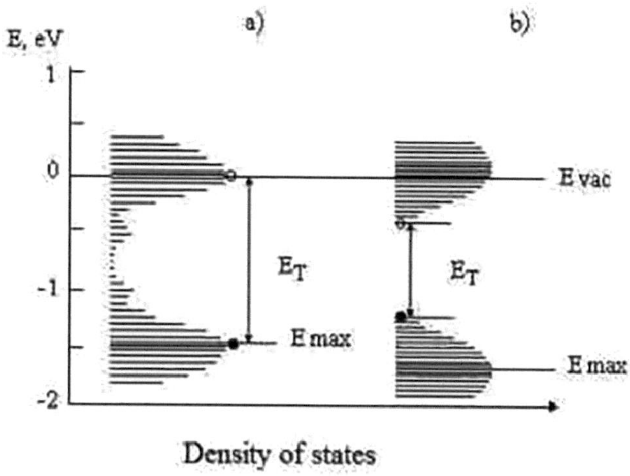

Figure 11. Schematic diagrams of the electronic structure of the optical absorption edge and optical gap of (a) insulating and transparent in the visible region and (b) absorbing and conductive a-C:H films.

The optical gap of the a-C:H films is determined by the energy of the most probable electronic π–π* transition. It should be noted a transition of a valence π-electron localized either on isolated linear chains or on complicated π-bound structural elements to a corresponding free π*-level in the conduction band. The spectra of these films contain weak discrete absorption bands, which lie below the energy E_{max} with peaks at ~1, ~0.7, and ~0.6 eV in [43]. They correspond to localized states within a pseudo gap whose distribution can also be described by a Gaussian curve. Scheme (b) of **Figure 9** describes the structure of the optical gap for a-C:H films with $E_{max} > E_T$ (samples *c* and *d* in **Table 2**). The discrete energy levels lying at the optical absorption edge below the level corresponding to the energy E_{max} form a dense quasi-continuous electron spectrum, which can be described by a Gaussian curve. There are essentially no states within the pseudo gap for $E_T \ll 1.0 \pm 0.1$ eV. In this case, E_T corresponds to the minimum transition energy of a valence π-electron to the corresponding empty π *-level in the conduction band. This kind of electronic structure is typical of a-C:H films, which absorb in the visible region and have a resistivity of $<10^9$ Ω cm. A low content of sp^3-bound states in the films ensures raising the conductivity. In the proposed schemes of **Figure 11**, the edge of the conduction band lies near the vacuum level

E_{vac} (scheme a) or below it (scheme b), consistent with current ideas regarding the electronic structure of a-C:H in [34]. The changes in the intensities of the absorption bands of the a-C:H were compared with changes in the resonance Raman scattering spectra to understand their electronic structure, the nature of the visible absorption, and their conductivity in [43].

5.2. Studying the a-C:H nanostructure using resonant Raman scattering spectroscopy

The Raman scattering spectra of amorphous carbon in its different modifications consisted of a broad band and can easily be resolved into two Gaussian-type bands. The first has a peak in the interval 1530–1580 cm^{-1} and was initially attributed to an active 1585 cm^{-1} line of single-crystal graphite, while the second band near 1300–1400 cm^{-1}, with a line at 1355 cm^{-1}, corresponds to a disordered mode in [46]. However, a high-frequency shifts of the principal maximum observed in the Raman scattering spectra of various modifications of a-C:H when the excitation energy is raised that has cast doubt on this interpretation in [47, 48]. It is the result of scattering from π-bond elements of the a-C:H structure, which is enhanced resonantly at photon energies approaching the π–π * resonance. The bands near 1400 and 1530 cm^{-1} may be associated with scattering on large- and small-sized π-bond clusters, respectively, in [48].

The absence of graphite clusters in the structure of fresh and thermally worked at a temperature of 400 °C in vacuum a-C:H films has been confirmed by comparing resonant Raman scattering spectra of a-C:H and graphite in [49]. The observed features of the a-C:H Raman scattering spectra lead to the conclusion that their structure includes a set of scattering centers characterized by different excitation energies of π–π * electronic transitions and vibration energies. The different sizes of the π-bond elements of the structure lead to different values of the coupling parameter. The strong disordered interaction among them causes the large spread in their electronic and vibration spectra. Resonant excitation of the Raman spectra produces a selective enhancement in the scattering at the frequencies of those centers for which these conditions are optimal. The width and shape of the Raman scattering bands are determined by the dependencies of the vibration frequency and location of an absorption band on the length of the coupling chain and on the size distribution of the elements. In this case, the procedure of expanding the complicated Raman scattering band and absorption bands in the electronic spectra into Gaussian profiles is arbitrary.

The way the relative intensity of the Raman scattering bands of fresh and thermally processed a-C:H depended on the excitation wavelength has been shown that these bands correspond to different types of structural elements, which are polyene chains of various lengths and polycyclic aromatic groups with different numbers of rings. The former predominates in the intensity of the band at ~1540 cm^{-1} and the latter, to that of the band at ~1340 cm^{-1}, is extremely probable, since in the Raman scattering spectra of the corresponding molecular structures, the most intense bands lie in the corresponding range of vibration frequencies in [49].

The existence of olefin chains in the structure is confirmed by studies of a-C:H films by elastic neutron scattering in [50] and by nuclear magnetic resonance and neutron diffraction in [51, 52]. Theoretical calculations of resonant Raman scattering spectra of amorphous carbon have shown that the band shapes are determined by a complex of sp^2-clusters with six or fewer rings. Polyene chains can contribute to the formation of the 1540 cm^{-1} band when the Raman scattering spectra are excited by visible light, while a lattice sp^3-structure has no direct effect on the resonant Raman scattering spectra in [53].

5.3. Comparison of Raman scattering and absorption spectra in the visible region

Resonant Raman scattering spectroscopy is an extremely informative and highly sensitive method for investigating the characteristics of middle-range π-bond elements in a-C:H structures. Increasing the energy of the ions involved in the condensation of a-C:H in RF in [54, 55], and DC glow discharge plasmas in [30] leads to a rise in the intensity ratio of the main bands at 1340 and 1540 cm^{-1} (I_{1340}/I_{1540}) in the Raman scattering spectrum. These changes can be explained by an increased amount of polycyclic aromatic groups compared to polyene chains in the a-C:H structure. Decreasing the amount of CH-groups in sp^3-states and the optical gap were observed in [42] concurrent with the rise in I_{1340}/I_{1540}. Since the Raman scattering spectra are caused by resonant scattering on π-bond elements of the structure, changes in these elements should correlate with changes in the electronic absorption spectra of a-C:H films. **Figure 12** shows plots of the band intensity ratio I_{600}/I_{800} in the absorption spectra and of the ratio I_{1340}/I_{1540} of the principal bands in Raman scattering spectra of a-C:H films obtained in [30, 49] vs. the voltage.

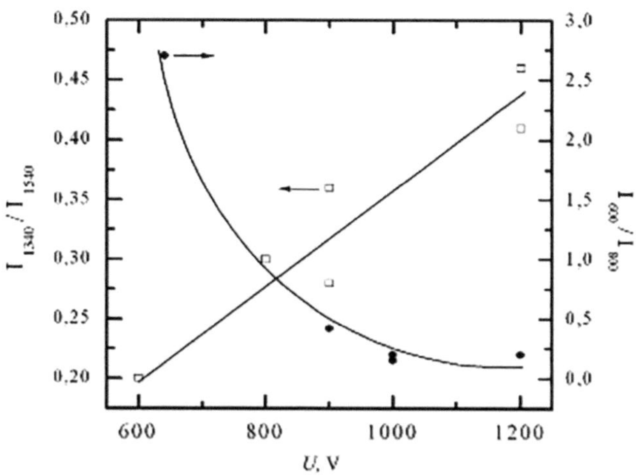

Figure 12. The variation in the ratio I_{600}/I_{800} of the integrated intensities of the principal bands in the absorption coefficient α (λ) and in the ratio I_{1340}/I_{1540} of the fundamental bands in the Raman spectra of a-C:H films vs. the voltage (U).

It can be seen that as the ratio I_{600}/I_{800} in the absorption spectrum decreases, the ratio I_{1340}/I_{1540} in the Raman scattering spectrum increases. It is known that the limiting wavelength in the absorption spectra of the polyene is 610 nm. Therefore, the peak near 600 nm in the absorption spectra correlates with the distribution of isolated polyene-type linear chains in the a-C:H structure. The Gaussian-like band at 800 ± 40 nm reflects the distribution of combined π-bond elements in the structure a-C:H. They can include polyene chains of various lengths, as well as polycyclic groups with a different number of aromatic rings. The many kinds of combinations of this π-bond structural elements with a single system for coupling of the multiple bonds, along with isolated polycyclic groups and polyene chains, are the most likely chromophores responsible for visible absorption in a-C:H films. Taking this circumstance into account, we may assume that the absorption spectra of a-C:H films are formed by the superposition of absorption spectra corresponding to electronic π–π^* transitions in isolated π-bond elements of the structure and they are quasi-continuous. The composition of the π-bond elements, their spatial distribution in the film structures, and the interaction probability of their π-electron systems determine the position of the peak, the intensity of the Gaussian-like bands which stand out in the spectrum, and the structure of the edge of the absorption band.

6. Application of optical properties of the films

The a-C:H films with $n \approx 2$ are ideal antireflection coatings for germanium with $n = 4$ in [22–24, 56, 57]. These films differ from well-known multilayer antireflection coatings in their high chemical durability and mechanical strength. The spectra in **Figure 13** illustrate the effect of the absorption in a-C:H films with $n = 2$ on the transmission of germanium. Single-layer coatings based on a-C:H films with the optical thickness 1/4 equal to 0.5 and 1.25 μm for $\lambda = 4$ and 10 μm, respectively, were deposited from acetylene by the above-described method onto both sides of a Ge substrate. Due to the absence of absorption in a-C:H in the range 3.5–5 μm, an increase of up to 100% in the transmission of Ge was obtained in this spectral region (**Figure 13**). For wavelengths above 5 μm, the effect of the absorption of IR radiation in the a-C:H films becomes more pronounced.

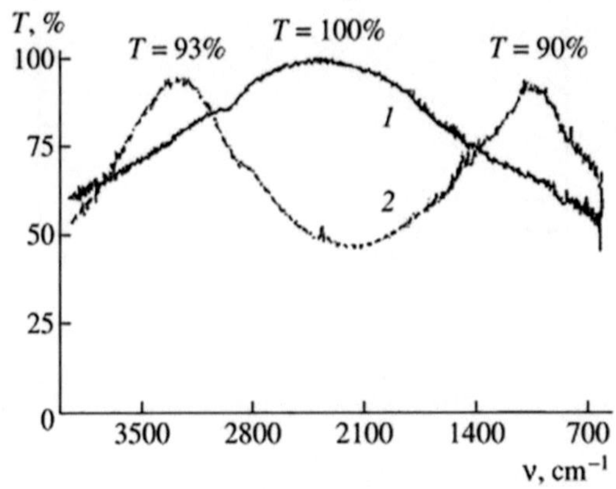

Figure 13. Transmission spectra (T) of germanium coated with quarter-wave a-C:H layers for wavelengths of (1) 4 and (2) 10 μm.

The transmission of a Ge sample that has both sides coated with a-C:H films does not exceed 90% at a wavelength of 10 μm due to the absorption loss in the a-C:H coatings in [27]. Due to their combination of chemical and mechanical durability, radiation resistance, and transparency in the IR region, diamond-like a-C:H films can be used as protective coatings for IR optics, in particular, for copper mirrors. Higher breakdown thresholds were observed on a-C:H coatings obtained with deposition rates <2 Å/s. Such coatings have the depths of trap levels not exceeding ~1.5 eV and by resistivity of 10^7–10^8 Ω cm in [58]. The a-C:H layer with the absorption coefficient about 4 × 10^4 cm^{-1}, and having resistivity in the interval from 10^{10} to 10^{11} Ω cm, and E_T in the interval from 1.25 up to 1.5 eV ensure effective optical isolation of the a-Si:C:H photoconductor in [44]. The intensity of reading out radiation has no influence on photo-addressing of the a-Si:C:H. The a-C:H films absorbing in the visible region were employed as a light-blocking layer in the optically addressed spatial light modulators operating in the reflective mode with a-Si:C:H photosensor. It is allowed to decide the problem of optical decoupling of the write and read light signals in [59–61]. The a-C:H films that are transparent in the visible spectral range, with n < 1.7 and high resistivity (~10^{13} Ω cm), can arbitrarily be referred to as polymer-like, which were used as orienting layers for liquid crystals. The orienting layers were obtained from liquid and gaseous hydrocarbons at low temperatures in [62].

7. Conclusion

Results of studies on the preparation of the a-C:H films by plasma-activated CVD in various hydrocarbon precursors and their structure and optical properties are reviewed. The dependences of the deposition rate of the films obtained by CVD process in a DC glow discharge with the magnetron plasma localized near the anode on the voltage, relative content of an inert gas in the mixture with a hydrocarbon, were discussed. Knowing them, one can control the process of condensation. One can vary the refractive index of a-C:H films in the interval 2.35–1.55 and modify their structure and properties from diamond-like carbon to polymer by increasing the deposition rate and the choice of the appropriate hydrocarbon precursor. Identifying the a-C:H film as a diamond-like or polymer-like film can use their vibration spectra, since they have specific features. A characteristic feature of the IR spectra of the diamond-like films is the presence of a band at ~1250 cm^{-1} due to the stretching vibrations of the C–C bonds in four-functional branching points of the structure. The spectra of polymer-like films show strong absorption bands due to the stretching and bending vibrations of CH groups, as well as of carbonyl and hydroxyl groups. The integrated intensity of the band of CH vibrations peaked at ~2900 cm^{-1} decrease exponentially by an order of magnitude with an increase in refractive indices from 1.55 to 2.4. Nanostructure of the a-C:H films can characterize using the absorption spectra in the visible region and the Raman scattering. Absorption in the range 400–600 nm is caused by π–π^* transitions in polyene chains, while the absorption in the range 600–800 nm is related to similar transitions in more complex π-bond combinations, including polycyclic aromatic groups. A conductivity a-CH films increase with decreasing a width of the optical gap. Diamond-like films can be used as an optical antireflection coating for infrared optical elements, and light-blocking layers in the visible spectral region. Transparent in the visible region polymer-like films can be used as alignment layer in LC devices due to their surface properties. A wide range of optical properties of the film holds great promise for application in different optical devices.

Author details

Elena A Konshina*

Address all correspondence to: eakonshina@mail.ru

ITMO University, Saint Petersburg, Russia

References

[1] Anderson IP, Berg S, Norstrom H, et al. Properties and coating rates of diamond-like carbon films produced de R.F. glow discharge of hydrocarbon. Thin Solid Films. 1979;63:155–157.

[2] Whitmel DS, Williamson R. The deposition of hard surface layers by hydrocarbon cracking. Thin Solid Films. 1976;35:255–261.

[3] Holland L, Ojha SM. Deposition of hard and insulating carbonaceous films on an r.f. target in a butane plasma. Thin Solid Films. 1979;58:107–116.

[4] Jones DI, Stewart AD. Properties of hydrogenated amorphous carbon films and the effects of dopping. Philos. Mag. B. 1982;45:432–434.

[5] Enke K, Dimigen H, Hubach H. Frictional properties of hard and insulating carbonaceous films grown in an R.F. discharge. Appl. Phys. Lett. 1980;36:291–292.

[6] Kim HT, Sohn SH. Effects of applied power on hydrogenated amorphous carbon (a-C:H) film deposition by low frequency (60 Hz) plasma-enhanced chemical vapor deposition (PECVD). Vacuum. 2012;86:2148–2151.

[7] McKenzie DR, McPhedran RC, Savvides N, Cockayne DJH. Analysis of films prepared by plasma polymerization of acetylene in D.C. magnetron. Thin Solid Films. 1983;108:247–256.

[8] Meyerson B, Smith FW. Electrical and optical properties of hydrogenated amorphous carbon films. J. Non-Cryst. Solids. 1980;35/36:435–440.

[9] Sung T-L, Chao Y-A, Liu C-M, Teii K, Teii S, Hsu C-Y. Deposition of amorphous hydrogenated carbon films on Si and PMMA by pulsed direct-current plasma CVD. Thin Solid Films. 2011;519:6688–6692.

[10] Mori T, Namba Y. Hard diamond-like carbon films deposited by ionized deposition of methane gas. J. Vac. Sci. Technol. A. 1983;1:23–27.

[11] Weismantel C, Bewilogua K, Bzeuer K, et al. Preparation and properties of hard i-C and i-BN coatings. Thin Solid Films. 1982;96:31–44.

[12] Hong J, Goullet A, Turban G. Ellipsometry and Raman study on hydrogenated amorphous carbon (a-C:H) deposited in a dual ECR-r.f. plasma. Thin Solid Films. 1999;352:41–48.

[13] Konshina EA. Deposition of a-C:H films in a DC glow discharge with the magnetron plasma localized near the anode. Tech. Phys. 2002;47:690–694.

[14] Konshina EA, Tolmachev VM, Vangonen AI, Fatkulina LA. Study of the properties of plasma-polymerized layers and their effect on the orientation of nematic liquid crystals J. Opt. Technol. 1997;64:476–482.

[15] Jacob W. Surface reactions during growth and erosion of hydrocarbon films. Thin Solid Films. 1998;326:1–42.

[16] Khan AA, Mathine D, Woollam IA. Optical properties of carbon films: an ellipsometric study. Phys. Rev.1983;28:7229–7235.

[17] Pascual E, Serra C, Esteve J, Bertran E. Ellipsometric study of diamond-like thin films. Surf. Coat. Technol. 1991;47:263–268.

[18] Serra C, Pascual E, Maass F, Esteve J. Plasma deposition of hydrogenated amorphous carbon (a-C:H) under a wide bias potential range. Surf. Coat. Technol. 1991;47:87–97.

[19] Babaev AA, Sultanov SB, Abdulvagabov MS, Terukov EI. Electrical, optical, and mechanical properties of amorphous hydrogenated carbon obtained under various deposition conditions. Semiconductors. 2011;45:118–120. doi:10.1134/S1063782611010039

[20] Hopf C, Angot T, Aréou E, Dürbeck T, Jacob W, Martin C, Pardanaud C, Roubin P, Schwarz-Selinger T. Characterization of temperature-induced changes in amorphous hydrogenated carbon thin films. Diam. Relat. Mater. 2013;37:97–103. doi:10.1016/j.diamond.2013.05.004

[21] Tolmachev VA, Konshina EA. Ellipsometric study of a-C:H films. Diam. Relat. Mater. 1996;5:1397–1401.

[22] Enke K. Hard carbon layer for wear protection and the antireflection purposes of infrared devices. Appl. Opt. 1985;24:508–512.

[23] Pellicori SF, Peterson CM, Henson TP. Transparent carbon films: comparison of properties between ion- and plasma-deposition processes. J. Vac. Sci. Technol. A. 1986;4:2350–2355.

[24] Klibanov L, Croitoru NI, Seidman A, Gilo V, Dahan R. Diamond-like carbon thin films as antireflective and protective coatings of GaAs elements and devices. Opt. Eng. 2000;39:989–992.

[25] Dischler B, Bubenzer A, Koidl P. Bonding in hydrogenated hard carbon studied by optical spectroscopy. Solid State Commun. 1983;48:105–108.

[26] Zhang F, Chen G, Zhang Y, Yu G. IR and ESR studies of the structural properties of hydrogenated amorphous carbon films. Phys. Status Solidi. (a) 1988;109:K39–K44.

[27] Konshina EA, Vangonen AI. Features of the vibrational spectra of diamond-like and polymer-like a-C:H films. Semiconductors. 2005;39:585–590.

[28] Memming R. Properties of a-C:H layers made by CVD. Thin Solid Films. 1986;143:279–289.

[29] Vangonen AI, Konshina EA, Tolmachev VA. Adsorption properties of the layers of substances inducing orientation of liquid crystals. Russ. J. Phys. Chem. A. 1997;71:981–985.

[30] Konshina EA, Baranov AV, Yakovlev VB. Vibration spectra of carbon films obtained from acetylene plasma. J. Appl. Spectr. 1988;48:624–628.

[31] Robertson J. Amorphous carbon. Adv. Phys. 1986;35:317–374.

[32] Robertson J, O'Reilly EP. Electronic and atomic structure of amorphous carbon. Phys. Rev. B. 1987;35:2946–2957.

[33] Robertson J. Properties of diamond-like carbon. Surf. Coat. Technol. 1992;50:185–203.

[34] Robertson J. Electronic structure of diamond-like carbon. D&RM. 1997;6:212–218.

[35] Dasgupta D, Demichelis F, Pirri CF, Tagliaferro A. π bands and gap states from optical absorption and electron-spin-resonance studies on amorphous carbon and amorphous hydrogenated carbon films. Phys. Rev. B. 1991;43:2131–2135.

[36] Dasgupta D, De Martino C, Demichelis F, Tagliaferro A. The role of π and π^* gaussian-like density-of-states bands in the interpretation of the physical properties of a-C and a-C:H films. J. Non-Cryst. Solids. 1993;164–166:1147–1150.

[37] Frainheim T, Blaudeck P, Stephan U, Jungnickel G. Atomic structure and physical properties of amorphous carbon and its hydrogenated analogs. Phys. Rev. B. 1993;48:4823–4834.

[38] Jungnickel G, Frainheim T, Porezag D, Blaudeck P, Stephan U, Newport RJ. Structural properties of amorphous carbon. IV. A molecular-dynamics investigation and comparison to experiments. Phys. Rev. B. 1994;50:6709–6716.

[39] Wanatabe I, Hasegawa S, Kurata Y. Photoluminiscense of hydrogenated amorphous carbon films. Jpn. J. Appl. Phys. 1982;22:856–859.

[40] Yamamoto K, Ichikawa Y, Nakayama T, Tawada Y. Relationship between plasma parameters and carbon atom coordination in a-C:H Films prepared by RF glow discharge decomposition. Jpn. J. Appl. Phys. 1988;27:1415–1421.

[41] Rohwer K, Hammer P, Thiele J-U, Gissler W, Blaudeck P, Frauenheim T, Meissner D. Electrical and optical properties of plasma-deposited amorphous hydrocarbon films. J. Non-Cryst. Solids. 1991;137–138:843–846.

[42] KonshinaE A. Correlation between the optical gap and features of the structure of amorphous hydrogenated carbon films. Phys. Solid State. 1995;37:610–612.

[43] Konshina EA. Absorption and optical gap of a-C:H films produced from acetylene plasma. Semiconductors. 1999;33:451–456.

[44] Tauc J, Grigorovici R, Vancu A. Optical properties and electronic structure of amorphous germanium. Phys. Status Solidi. (b) 1966;15:627–637.

[45] Konshina EA, Feoktistov NA. Properties of a-C:H light-blocking layer used for optical isolation of a-SI:C:H photosensor in reflective liquid-crystal spatial light modulators. J. Phys. D. Appl. Phys. 2001;34:1131–1136.

[46] Dillon RO, John A, Wollam JA, Katkanant V. Use of Raman scattering to investigate disorder and crystallite formation in as-deposited and annealed carbon films. Phys. Rev. B. 1984;29:3482–3489.

[47] Ramsteiner M, Wagner J. Resonant Ramon scattering of hydrogenated amorphous carbon: evidence for π-bonded carbon clusters. Appl. Phys. Lett. 1987;51:1355–1357.

[48] Yoshikawa M, Katagiri G, Ishida A, Ishitani H, Akamatsu T. Resonant Raman scattering of diamondlike amorphous carbon films. Appl. Phys. Lett. 1988;52:1639–1641.

[49] Baranov AV, Konshina EA. Resonance Raman-scattering of hydrocarbon films deposited from acetylene plasma. Opt. Spectr. 1988;65:506–509.

[50] Walters JK, Honeybon P, Huxley DW, Newport RJ, Howells WS. Structural properties of hydrogenated carbon. I. A high-resolution neutron-diffraction study. Phys. Rev. B. 1994;50:831–845.

[51] Jager M, Gottward J, Spiess HW, Newport RJ. Structural properties of amorphous hydrogenated carbon.III. NMR Investigations. Phys. Rev. B. 1994;50:848–852.

[52] Honeybone JR, Newport RJ, Walters JK, Howells WS, Tomkinson J. Structural properties of amorphous hydrogenated carbon. II. An inelastic neutron-scattering study. Phys. Rev. B. 1994;50:839–845.

[53] Matsunuma S. Theoretical simulation of resonance Raman bands of amorphous carbon. Thin Solid Films. 1997;306:17–22.

[54] Tamor MA, Haire JA, Wu CH, Hass KC. Correlation of the optical gaps and Raman spectra of hydrogenated amorphous carbon films. Appl. Phys. Lett. 1989;54:123–125.

[55] Xu S, Hundhausen M, Ristein J, Xan B, Le L. Influence of substrate bias on the properties of a-C:H films prepared by plasma CVD. J. Non-Cryst. Solids. 1993;164–166:1127–1130.

[56] Zelez J. Diamond-like carbon films. RCA Rev. 1982;43:665–673.

[57] Bubenser A, Dischler B, Brandt G, Koidl P. Role of hard carbon in the field of infrared coating materials. J. Appl. Phys. 1983;54:4590–4595.

[58] Konshina EA. Interaction of intense IR laser radiation with a-C:H protective coatings. Tech. Phys. 1998;43:1062–1068.

[59] Konshina EA, Onokhov AP. Employment of absorbing a-C:H films in reflective liquid-crystal light modulators. Tech. Phys. 1999;44:298–299.

[60] Onokhov AP, Konshina EA, Feoktistov NA, Beresnev LA, Haase W. Reflective type FLC optically addressed spatial light modulators with pixelated metal mirror and light-blocking layer. Ferroelectrics. 2000;246:259–68.

[61] Berenberg VA, Venediktov VY, Ivanova NL, Isaev MV, Konshina EA, Onokhov AP, Fedorov MA, Feoktistov NA. Large-aperture optically addressed spatial light modulator development. In: Proceeding SPIE of XV International Symposium on Gas Flow, Chemical Lasers, and High-Power Lasers., Bellingham, WA. 2005. V. 5777. (PART II), pp. 711–715. doi:10.1117/12.611249

[62] Konshina EA. Production methods and properties of liquid-crystal-orienting layers based on amorphous carbon. J. Opt. Technol. 2011;78:210–217. doi:10.1364/JOT.78.00021

Chapter 8

Amorphous, Polymorphous, and Microcrystalline Silicon Thin Films Deposited by Plasma at Low Temperatures

Mario Moreno, Roberto Ambrosio, Arturo Torres,
Alfonso Torres, Pedro Rosales, Adrián Itzmoyotl and
Miguel Domínguez

Additional information is available at the end of the chapter

http://dx.doi.org/10.5772/63522

Abstract

The present chapter is devoted to the study of amorphous (a-Si:H), polymorphous (pm-Si:H), and microcrystalline (μc-Si:H) silicon, deposited by the plasma-enhanced chemical vapor deposition (PECVD) technique at low temperatures. We have studied the main deposition parameters that have strong influence on the optical, electrical, and structural properties of the polymorphous and microcrystalline materials. Our results reveal the key deposition conditions for obtained films with optical and electrical characteristics, which are suitable for applications on thin-film solar cells and semiconductor devices.

Keywords: amorphous, polymorphous, microcrystalline, silicon, nanocrystals

1. Introduction

At the present time hydrogenated amorphous silicon (a-Si:H) is a mature material of the microelectronics and photovoltaic industries. Its success is due to the compatibility with the silicon CMOS technology, the possibility of doping (n or p type), the low substrate temperatures used (≤300°C) when is deposited by the plasma-enhanced chemical vapor deposition (PECVD) technique, and the possibility to deposit it over very large substrate areas (>1 m^2), such as glass, metal foils, or plastic flexible substrates. Thin-film transistors (TFTs), large-area thin-film solar cells, and high-performance infrared (IR) cameras (based on microbolometer arrays) among others are devices based on the a-Si:H technology.

However, despite the important characteristics of a-Si:H, it has several drawbacks, such as a large density of defects, poor transport properties such as low carriers' mobility, and poor stability against radiation [1, 2], which limits its applications in new and high-performance devices. In this aspect, there exists a constant study of the optimization of the deposition conditions of a-Si:H films by the PECVD technique, in order to improve its performance characteristics and to produce other materials with different structural, electrical, and optical characteristics.

Recently, it has been demonstrated that it is possible to produce nanocrystals, of about 3–5 nm of diameter, in the a-Si:H matrix, by modifying the deposition conditions of standard a-Si:H, using the PECVD technique. The above material is commonly referred as polymorphous silicon (pm-Si:H) [3–6]; the presence of nanocrystals distributed on the silicon amorphous matrix reduces the density of states (DOS) and defects and improves the electrical properties, as carriers' mobility and stability of the films against radiation [3–6]. Moreover, pm-Si:H also preserves the characteristics of a-Si:H as a direct band gap, high absorption coefficient, and high activation energy (E_a).

Therefore pm-Si:H films can replace a-Si:H in devices as the intrinsic film in a-Si:H thin-film solar cells. It has been reported that a-Si:H thin-film solar cells suffer from degradation due to constant illumination (Staebler-Wronski effect, SWE [7]); therefore the initial solar cell efficiency is reduced. As well it has been demonstrated that the use of pm-Si:H instead of a-Si:H reduces the light-induced degradation [8].

On the other hand, also by modifying the silicon film deposition conditions using PECVD, it is possible to increase the crystal sizes and the crystalline fraction of the silicon films and therefore modifying not just the film structure but also the electro-optical characteristics, as the absorption coefficient, the room temperature conductivity, the band gap, and E_a. Micro-crystalline silicon (μc-Si:H) is a silicon film where the sizes of the silicon crystals are much more larger than in pm-Si:H films, with sizes in the order of dozens of nanometers, and the overall crystalline fraction (X_C) is large, and the electro-optical characteristics are different than its a-Si:H/pm-Si:H counterpart.

The μc-Si:H thin films are suitable for various devices, such as TFTs, due their larger room temperature conductivity and carrier mobility (compared to those of a-Si:H or pm-Si:H), which are transduced in faster devices, as well the band gap and absorption coefficient are different than those of a-Si:H, resulting in thin solar cells with larger IR absorption and more stable against sun radiation. At the present time a-Si:H/μc-Si:H tandem solar cells (or also known as micromorph solar cells) have been developed with stabilized efficiencies of about 12% [9].

2. Plasma deposition

The a-Si:H films used in the industry are typically obtained by means of PECVD. The first work on the deposition of a-Si:H material by silane glow discharge started in 1960s [10].

The standard PECVD system consists of a reactor of two parallel electrodes capacitively coupled, separated at some distance. The substrate is placed in one of these electrodes, as is shown in **Figure 1**. The most common RF is 13.56 MHz. In the PECVD reactor, silane (SiH_4), often diluted with other precursor gases (H_2, Ar, etc.), is pumped at a certain flow rate. A substrate for the film deposition is collocated in the chamber in the bottom electrode, and a RF oscillating voltage is applied to the upper electrode of the chamber. Most of the surface chemistry occurs due to the high-energy electron bombardment. The deposition process of silicon-based films can be described as a main four-step process:

1. The primary reaction is made by the glow discharge, which results in electron-impact excitation, dissociation, and ionization of the molecules. The plasma thus consists of neutral radicals and molecules, positive and negative ions, and electrons.
2. Reactive neutral species are transported to the substrate by diffusion, positive ions bombard the growing film, and negative ions are trapped within the sheaths and may eventually form small particles.
3. The third step consists of surface reactions, such as hydrogen abstraction, radical diffusion, and chemical bonding. Then the formation of islands is made and it continues until there is a thin film.
4. The fourth step is the subsurface release of hydrogen molecules from the film.

The deposition process involved physical and chemical interactions in the plasma and at the growing film surface is dependent on several parameters such as RF power, frequency, substrate temperature, gas pressure, and electrode geometry; a deeper explanation could be found in [11].

Generally a plasma deposition system consists of several subsystems, each providing different functions. The reactor system is the central part where the molecules are dissociated and the products are deposited on heated substrates in order to form a layer. The reactor has a capacitor electrode configuration. The power to the reactor system is delivered by means of the RF source connected via the matching network. The RF source generally operates at a frequency of 13.56 MHz. The power and ground are connected to the top and bottom electrodes, respectively. The samples are loaded in the bottom electrode, which can be heated from room temperature to about 300°C. The gas control system includes mass flow controllers to measure and control the different gases (GeH_4, SiH_4, NH_3, Argon, H_2, PH_3, B_2H_3) supplied to the chamber. The vacuum system comprises mechanical pump, turbo molecular pump, roots pump, and the pressure controller system. The deposition pressure can be controlled in a wide range of values, usually from 500 to 2000 mTorr.

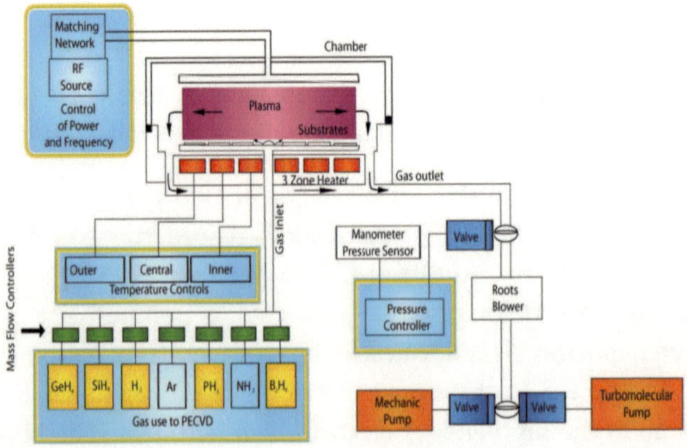

Figure 1. Scheme of a plasma deposition system (PECVD) used for film deposition. Also the power, frequency, temperature, gas flow, and pressure controls are shown.

In our particular facilities, we use a four-chamber cluster tool from MVSystems for the deposition of a-Si:H and related materials. To minimize cross-contamination between the chambers, it has a chamber for the deposition of intrinsic a-Si:H films, two chambers for the deposition of p- and n-doped semiconductor films, and a chamber for the deposition of metals and transparent conductive oxides, as is shown in **Figure 2**. The chambers are kept under high vacuum (10^{-6} torr) by turbo molecular pumps. The cluster tools eliminate cross-contamination in multilayer thin-film structures and allow the production of high-quality electronic devices [12].

Currently, several deposition techniques have been reported for the deposition of a-Si:H and nano- and microcrystalline materials, such as radio-frequency (13.56 MHz) capacitively coupled PECVD (RF-PECVD) [11], very high-frequency PECVD (VHF-PECVD) [13], microwave CVD (MW-CVD) [14], electron cyclotron resonance CVD (ECR-CVD) [15], reactive magnetron sputtering (RMS) [16], hot-wire CVD (HW-CVD) [17], and the remote expanding thermal plasma (ETP) [18].

Figure 2. PECVD cluster tool from MVSystems, at the facilities of the Laboratory of Microelectronics of INAOE (project no. 152224 of the Secretary of Energy, SENER and the National Council of Science and Technology, CONACYT, Mexico).

http://dx.doi.org/10.5772/63522

One of the drawbacks for RF-PECVD at 13.56 MHz is the deposition rate, which is in the range of 1–3 Å/s. However, some efforts of increasing the RF-PECVD growth rate using higher power and pressure have been reported in [19]. The best deposition rates using the VHF in the range from 30 to 300 MHz have been reported until 20 Å/s [20].

3. Amorphous silicon

The noncrystalline semiconductors materials also known as "amorphous semiconductors" are usually obtained by the dissociation of gas species using the PECVD technique [21, 22]. In these materials the chemical bonding of atoms is a random covalent network; the disorder variation in the angles between bonds eliminates the regular lattice structure of its crystalline counterpart, as is shown in **Figure 3**. However, the noncrystalline semiconductors have demonstrated good optical and electronic properties for many device applications.

Hydrogenated a-Si:H is a mature material in the electronics manufacturing industry, where it is used for the development of thin-film solar cells, IR sensors, and TFTs, due to its very large absorption coefficient, its compatibility with the standard Si-CMOS technology, and the low process temperatures used (<300°C) when it is deposited by PECVD. The above features allow the use of different substrates such as metal foils, plastics, and glasses. In **Table 1** a comparison of Si-CMOS technology in contrast to the TFT technologies is shown. From the point of view of the process temperature and cost, a-Si:H technology is very attractive for flexible and low-cost electronics.

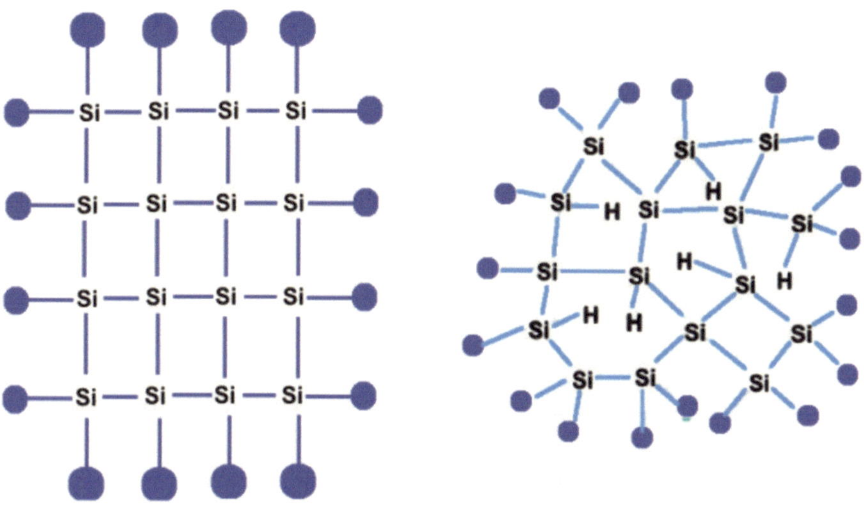

a) Crystalline structure b) Amorphous structure

Figure 3. Schematic representation for the structures of (a) crystalline silicon (c-Si) and (b) amorphous silicon (a-Si:H).

The a-Si:H film is composed by covalent bonds of Si—Si and Si—H. The silicon atoms form covalent bonds mainly with three neighboring silicon atoms with a hydrogen atom [23]. The hydrogen contents in a-Si:H have been reported from 4% up to 40%, and the density of hydrogen atoms depends on deposition conditions. The amorphous structure of the material leads to open network structures, in which voids and un-terminated bonds are so-called defects or dangling bonds. For integration of a-Si:H into electronic devices, the defect density must be in the range from 10^{15} to 10^{16} cm^{-3} [24]. In pure a-Si, the defect density is higher (10^{19} to 10^{20} cm^{-3}) than in hydrogenated a-Si:H, which is not suitable for electronic device applications. In a-Si:H the hydrogen is responsible for the lowering of the defect density by passivation of these dangling bonds [25]. This is crucial because of the influence of the optoelectronic properties, since in a-Si:H the dangling bonds can act as efficient recombination centers for electrons and holes. The increase in defect density with light radiation (light soaking) is the main cause of the SWE [26], where the light induces a creation of metastable defects in a-Si:H. These additional dangling bonds can be reduced by heating the sample up to approximately 200°C. Some works have reported the reduction of that effect by incorporating fluorine in the gas mixture during production [27].

Parameter Technology	Device technology		
	c-Si MOSFET	**a-Si:H TFT**	**Organic TFT**
Process temperature	1000°C	200°C	<100°C
Process technology	Photolithography multilayers	Photolithography multilayers	Shadow mask and ink-jet
Design rule	<32 nm	8 μm	50 μm
Substrate	Wafer	Glass/plastic	Plastic/metal foil
Supply voltage	1 V	20 V	Ink-jet 40 V
Device type	N-type and P-type	N-type	P-type
Mobility	1500 cm^2/Vs	1 cm^2/Vs	0.5 cm^2/Vs
Cost/area	High	Medium	Low
Lifetime	Very high	High	Medium–low

Table 1. Comparison of TFT and CMOS technology [23].

In amorphous semiconductors, the disorder and the presence of dangling bonds have an influence on the electronic DOS. **Figure 4** shows a simple schematic diagram of the DOS of the conduction band (delocalized states) and the valence band (delocalized states) from the Davis-Mott model, where E_c and E_v represent the energies, which separate the ranges where the states are localized and extended. The central band is split into a donor and an acceptor band.

In the above model, the mobility has a drop in several orders of magnitude at the transition from extended to localized states, resulting on a mobility edge. Thus the interval between E_c and E_v acts as a pseudo-gap (referred as the mobility gap) [28].

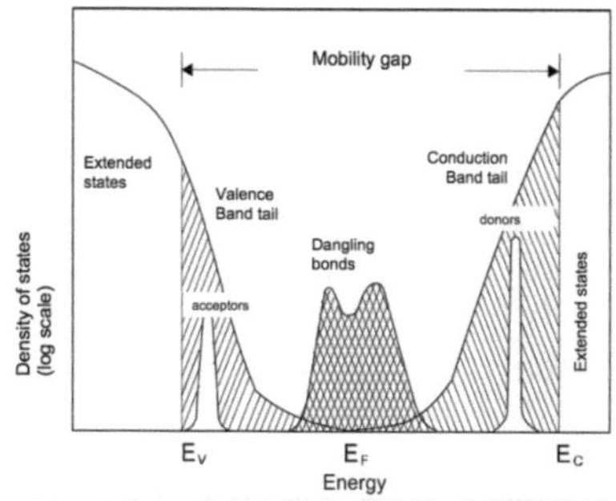

Figure 4. Schematic representation of DOS based on Davis-Mott model [28].

There are several definitions for optical gap and conventionally Tauc optical gap is used among others. Thus, the band gap can be approximated using the measurements of the optical absorption coefficient.

$$\alpha(h\nu)=(A/h\nu)(h\nu-E_T)^2 \tag{1}$$

where h is Planck's constant, α is the absorption coefficient, ν is the radiation frequency, E_T is the optical band gap, and A is a proportional constant.

The absorption coefficient of a-Si:H is showed in **Figure 5**, where three main regions in the spectra are typical identified, which are due to (a) band-to-band transitions that correspond to the transitions from valence to conduction band, (b) transitions related to valence and conduction band tails, and (c) defect-related transitions.

Figure 6 shows a HRTEM cross-sectional view of an a-Si:H thin film deposited in our facilities; in the figure it is observed that there is not a larger range order of the atoms in the material, which is the main structural characteristic of amorphous semiconductors.

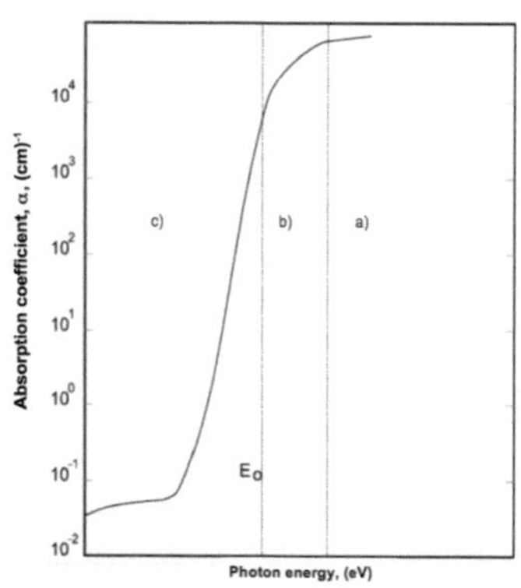

Figure 5. Typical absorption coefficient of a-Si:H plotted as a function of photon energy [28].

Figure 6. Cross section of an a-Si:H thin film observed by HRTEM.

4. Polymorphous silicon (pm-Si:H)

Basically pm-Si:H is a-Si:H with embedded nanocrystals of about 3–5 nm of diameter; those nanocrystals are distributed in the a-Si:H matrix. The pm-Si:H films are obtained by modifying the deposition conditions of standard a-Si:H, using the PECVD technique. The presence of nanocrystals on the silicon amorphous matrix reduces the DOS and defects and improves the electrical properties, as the carriers' mobility and the stability of the films against radiation [3–6]. Also, pm-Si:H preserves the characteristics of a-Si:H as a direct band gap, high absorption coefficient, and high activation energy (E_a). Intrinsic pm-Si:H has a direct optical band gap (1.6–1.8 eV) and a very large activation energy ($E_a \approx 1$ eV), when it is deposited by PECVD.

The above properties of pm-Si:H make this material very suitable for more stable devices as thin-film solar cells and thermal detectors. In this aspect, recently it has been reported thin-film solar cells where the a-Si:H intrinsic film has been replaced with a pm-Si:H film, resulting on solar cells with higher stability against light radiation (light soaking), in comparison with its counterpart employing a-Si:H [8].

4.1. Deposition conditions of pm-Si:H

We have performed a study of the deposition conditions of pm-Si:H by PECVD [21]. The deposition parameters were varied in order to observe their effect on the structural, electric, and optical characteristics of the films deposited. A series of pm-Si:H thin films was deposited by PECVD at RF = 13.56 MHz, from a SiH_4 and H_2 mixture. **Table 2** shows the deposition conditions for the pm-Si:H thin films; those films were deposited for 30 min, at substrate temperature $T_s = 200°C$. For the pm-Si:H film deposition, a high H_2 dilution is used and the chamber pressure was varied, since it is an important parameter for the production of nanocrystals. Notice that here we present some selected processes, which are more representatives.

Sample #	Press. (mTorr)	Power (W)	Power density (mW/cm²)	SiH_4 flow (sccm)	H_2 flow (sccm)
P1	500	20	69	50	50
P2	1000	20	69	50	50
P3	1500	20	69	50	50

Table 2. Deposition conditions of pm-Si:H films.

4.2. Characterization of pm-Si:H

4.2.1. Deposition rate

The deposition rate (V_d) of the pm-Si:H films was calculated from the average thickness and the deposition time; in this case all the films were deposited for 30 min. **Figure 7** shows the variation of deposition rate of the pm-Si:H films as a function of deposition pressure, maintaining constant the other process parameters (see **Table 1**). In **Figure 7**, it is seen that deposition rate of the pm-Si:H films increases from 0.5 Å/s to 2.4 Å/s as the deposition pressure increases. For the pressure of 0.5 torr, the deposition rate of the film is 0.5 Å/s (which is a pressure generally used for the deposition of a-Si:H). At a pressure of 1.5 torr, the deposition rate increases to about 2.4 Å/s.

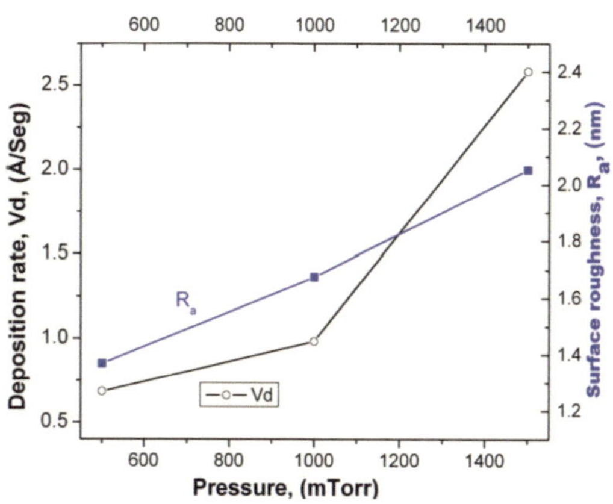

Figure 7. Deposition rate and surface roughness of pm-Si:H films as a function of the deposition pressure.

4.2.2. Structural analysis

Figure 8(a)–(c) shows three-dimensional (3D) AFM surface images of the pm-Si:H films deposited on silicon substrates; those films were deposited at different pressure values: 500, 1000, and 1500 mTorr. In the figure it is observed that the film deposited at 1500 mTorr has the largest R_a (of about 2.05 nm), which is in agreement with the deposition rate V_d results, since larger V_d and larger roughness are associated with larger crystallinity fraction in the films. The dependence with the deposition pressure of the average surface roughness (R_a) is shown in **Figure 7** and those values are listed in **Table 3**.

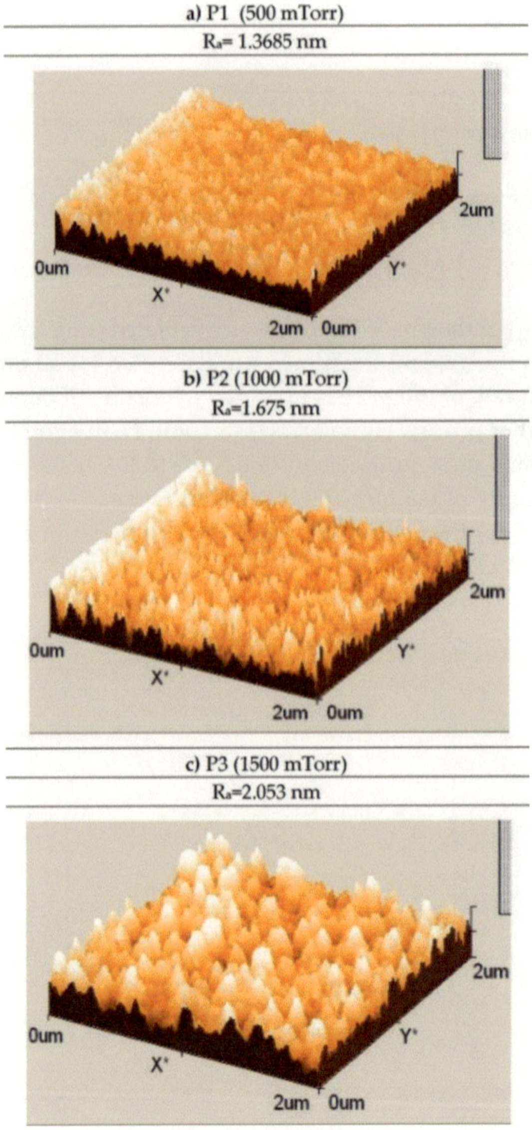

Figure 8. AFM 3D images of pm-Si:H films as a function of the deposition pressure: (a) P1 (500 mTorr); R_a = 1.3685 nm. (b) P2 (1000 mTorr); R_a = 1.675 nm. (c) P3 (1500 mTorr); R_a = 2.053 nm.

High-resolution transmission electron microscopy (HRTEM) was used to study the film's bulk transversal structure. **Figure 9** shows a transversal view of a 100 nm thick pm-Si:H film deposited over (1 0 0) c-Si obtained in our previous work [22]; in the figure several nanocrystals of sizes of around (3–5 nm) are marked; it also is possible to observe the presence of nano-islands at the c-Si/pm-Si:H interface.

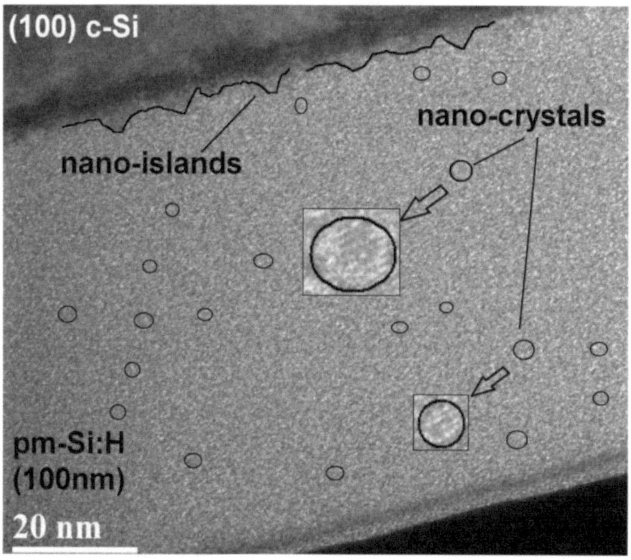

Figure 9. HRTEM image of the transversal view of a pm-Si:H film deposited by PECVD [22].

Sample #	R_a (nm)	C_H (%)	$X_{C\_(Raman)}$ (%)
P1	1.36	2.3	No sensitivity
P2	1.67	9.9	No sensitivity
P3	2.05	15.9	No sensitivity

Table 3. Surface roughness (R_a), H_2 content, and crystalline volume fraction for the pm-Si:H films.

4.2.3. Fourier transform infrared spectroscopy (FTIR) analysis

The IR absorbance spectra of the pm-Si:H thin films in the range 500–2200 cm^{-1} are shown in **Figure 10**. In the spectra several peaks related to different vibration modes are found. The peak at around 640 cm^{-1} corresponds to Si—H rocking/waging modes [29], the band at around 800–950 cm^{-1} is related to SiH_2 or SiH_3 bonding with low intensity, and it depends on deposition conditions [29–31]. The band at 1900–2100 cm^{-1} is related to SiH, SiH_2, and SiH_3 stretching modes.

In **Figure 10**, it is observed that as the pressure deposition increases, a change in the shape and intensity of the band at 640 cm^{-1} is observed. For samples, P1, P2, and P3, the total H_2 bonded content C_H of the films was calculated with the integration of the area under the Si—H mode (at 640 cm^{-1}) using Eq. (2) [29]:

$$C_H(\%) = \frac{A_\omega}{N_{Si}} \int \frac{\alpha(\omega)}{\omega} d\omega \quad (2)$$

where α is the absorption coefficient at frequency (ω), A_ω is the oscillator strength ($A_{(640)} = 1.6 \times 10^{19}$ cm^{-2}) [29, 32], and $N_{Si} = 5 \times 10^{22}$ cm^{-3} is the atomic density of pure silicon. The IR spectrum was deconvoluted using Gaussian peaks centered for the region of interest. **Table 3** shows the H_2 content results and is observed that H_2 increases as the pressure increases. Also in **Figure 10** a shift from 700 cm^{-1} to 640 cm^{-1} from the film deposited at lower pressure to the film deposited at higher pressure is observed, which could be related to the evolution of hydrogen content on the films, and associated to the presence of nanocrystals in the films deposited at higher pressure [33]. At higher pressures there is a change in the discharge regime, which results in the formation of silicon nanocrystals (up to 2 nm), during the growth of pm-Si:H films [34, 35], and that is in agreement with the values of R_a in our AFM analysis.

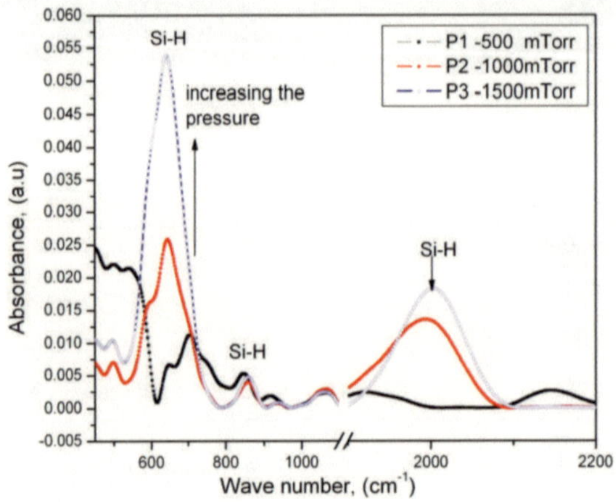

Figure 10. FTIR spectra of the pm-Si:H thin films.

4.2.4. Raman analysis

Raman-scattering measurements were performed in the pm-Si:H films. Raman spectra were deconvoluted into three bands related to an amorphous phase (480 cm^{-1}), an intermediate phase associated with small-size nanocrystals (500–514 cm^{-1}), and a crystalline phase (520 cm^{-1}). **Figure 11** shows the Raman spectra of the pm-Si:H film series.

The Raman spectra of the pm-Si:H films have a main peak located at 479 cm^{-1} (corresponding to a transverse optical phonon), which is related to an amorphous phase, and there was not detected any phase related to a crystalline fraction (520 cm^{-1}). The contribution of the nano-crystals in pm-Si:H films is too small in order to be detected by Raman scattering.

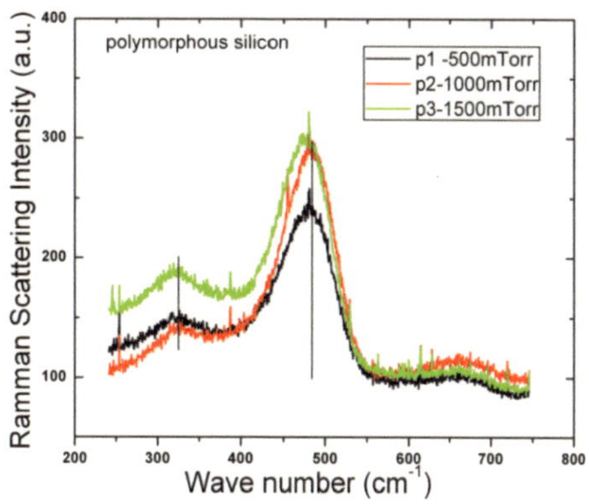

Figure 11. Raman spectra of the pm-Si:H thin films.

4.2.5. Optical properties

The optical band gap (E_{opt}) of pm-Si:H films was calculated from transmittance measurements and the method of Tauc. In **Figure 12** the Tauc plots of pm-Si:H films are shown, where the E_{opt} values were extracted. In **Table 4** the E_{opt} values are shown, and it is observed that with an increase of deposition pressure, E_{opt} has a decrement, from 1.93 eV to 1.84 eV. The obtained values in the pm-Si:H series for E_{opt} are in agreement for intrinsic pm-Si:H films which have a direct E_{opt} in the range of 1.6–1.8 eV and very large activation energy (E_a = 1 eV), when they are deposited by PECVD [35–37].

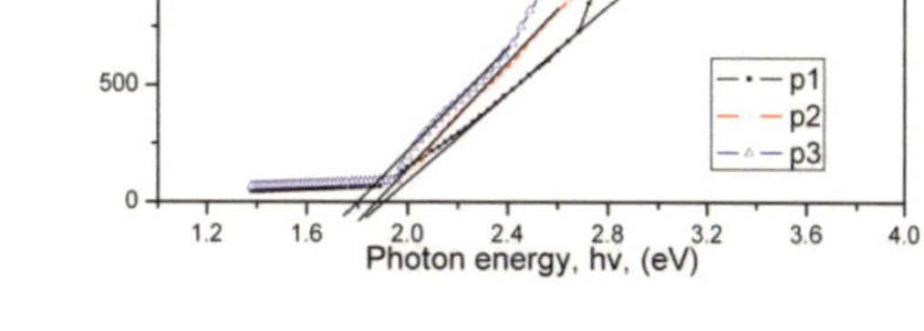

Figure 12. Optical band gap determination from the Tauc plots ($(\alpha h\nu)^{1/2}$ vs. photon energy) for pm-Si:H films as a function of the deposition pressure.

4.2.6. Electrical properties

Temperature dependence of conductivity (σ(T)) measurements were performed in the pm-Si:H films; from that characterization, the activation energy E_a was extracted; those values are shown in **Table 4**, where it is observed that the E_a values are in the range of 0.72–0.94 eV, which is in agreement with those values reported in literature [34].

As well the dark room temperature conductivity (σ_{dark}) and the photoconductivity (σ_{photo}) under AM 1.5 illumination conditions were measured in the pm-Si:H films. Notice that in the films deposited at larger deposition pressure, the σ_{dark} tends to increase as well (**Table 4**). The larger increment of σ from dark to AM 1.5 illumination in the pm-Si:H films is close to six orders of magnitude, measured in the film deposited at 1500 mTorr, which is a very interesting result for thin-film solar cells.

Sample #	Press. (mTorr)	V_d (A/s)	E_{opt} (eV)	σ_{dark} (Ω-cm)$^{-1}$	σ_{ph} (Ω-cm)$^{-1}$	E_a (eV)
P1	500	0.55	1.93	damaged	damaged	damaged
P2	1000	0.7	1.88	6.23E–9	1.04E–5	0.72
P3	1500	2.4	1.84	1.39E–9	1.01E–3	0.94

Table 4. Characteristics of pm-Si:H films deposited.

5. Microcrystalline silicon (μc-Si:H)

μc-Si:H is a silicon-based thin film, with crystals in the range of 20–700 nm [38]. Those crystals are of different orientations and grown in columns which are separated by an amorphous phase. Usually μc-Si:H films are grown from SiH_4 and H_2 gas mixtures, but also SiF_4, H_2, and Ar mixtures have been used [39]. The main parameters to grow μc-Si:H films are high H_2 dilution, moderated RF power, and high deposition pressure; according to the optimization of these parameters, the crystalline fraction can increase (Xc) and also its performance characteristics can be optimized.

In the last years μc-Si:H has gained attention due to its properties which are similar than those of polycrystalline silicon deposited by the low-pressure CVD (LPCVD). Polycrystalline silicon has a very large carriers' mobility and high stability, which make it very suitable for TFTs; however high deposition temperatures are needed (400–600°C), which limits its use in flexible substrates and even on glasses. On the other hand, μc-Si:H is deposited at low temperatures (200°C) and also exhibits high carriers' mobility, high stability, and high conductivity.

For thin-film solar cell applications, μc-Si:H also has been extensively studied due to its larger IR absorption than that of a-Si:H [39] and also larger stability against sun radiation (light soaking). At the present time, a-Si:H/μc-Si:H tandem solar cells (micromorph solar cells) have been developed with stabilized efficiencies up to 12% [9].

5.1. Deposition conditions of μc-Si:H

We performed a study for the production of μc-Si:H thin films in a PECVD reactor. The main parameters for the production of μc-Si:H are high pressure, moderated power, and high H_2 dilution. A series of μc-Si:H was deposited by PECVD at RF = 13.56 MHz, from a SiH_4 and H_2 and Ar mixture. **Table 5** shows the deposition conditions for the μc-Si:H thin films; those films were deposited for 30 min at substrate temperature T_s = 200°C. For the growth of the μc-Si:H films, Ar dilution has been used and the effect of the variation of the RF power density has been studied. Notice that here we present selected processes, which are more representa-tives.

Sample #	Pressure (mTorr)	Power (W)	Power density (mW/cm²)	SiH_4 flow (sccm)	H_2 flow (sccm)	Ar flow
M6	1500	25	83	10	41	Yes
M7	1500	30	103	10	41	Yes
M8	1500	35	121	10	41	Yes

Table 5. Deposition conditions of μc-Si:H films.

5.2. Characterization of μc-Si:H

5.2.1. Deposition rate

The deposition rate (V_d) of the μc-Si:H films was calculated from the average thickness and the deposition time; notice that all the films were deposited for 30 min. **Figure 13** shows V_d as a function of the RF power used for the deposition of the films.

In literature it has been found that the transition from amorphous to μc-Si:H, using SiH_4 and Ar mixtures, is produced by an increment of the RF power density [37, 40]. In **Figure 13** is observed that the deposition rate decreases as the RF power increases, from 1 Å/s to 0.8 Å/s; however those values of V_d are closed to those of the films reported in literature [37, 41].

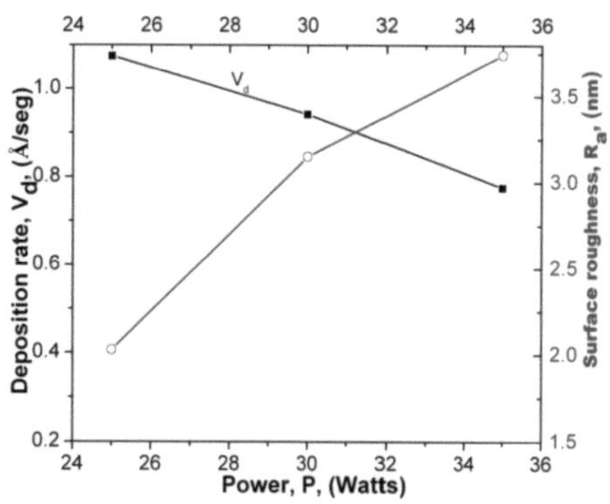

Figure 13. Deposition rate and surface roughness of μm-Si:H films as a function of the deposition pressure.

5.2.2. Structural analysis

Figures 14 (a)–(c) shows 3D AFM surface images of μc-Si:H films. In the μc-Si:H films where the RF power density was varied, we found that the film deposited at largest RF power (35 W) had the largest R_a value. The values of R_a for the μc-Si:H films are in the range of 1.9–2.5 nm (**Table 6**). Larger RF power used in the film deposition resulted in an increment of the film surface roughness, which is in agreement with [42], since larger R_a is observed in films with larger crystalline fraction [35].

Figure 15 shows a top image of a μc-Si:H thin film obtained by ultrahigh-resolution field emission scanning electron microscope (FEI Scios Dual Beam); in the figure grains of diameters of about 50 nm are observed at the surface of the μc-Si:H thin film. The surface structure is in agreement with those results observed by AFM.

Despite AFM is used to correlate the presence of silicon clusters in the film surface with the presence of silicon nanocrystals in the bulk of the film, by itself this technique is not sufficient to determine the structural composition in the bulk of the film.

HRTEM and Raman spectroscopy are techniques more suitable to determine the microcrystalline nature of the films. Figures 16 and **17** show a HRTEM transversal view of a μc-Si:H film (obtained with a FEI HRTEM Thalos F200X). In **Figure 16** crystals with different orientations can be observed, while in **Figure 17**, an amplified image is shown, where the atomic order in crystals of different orientations is observed.

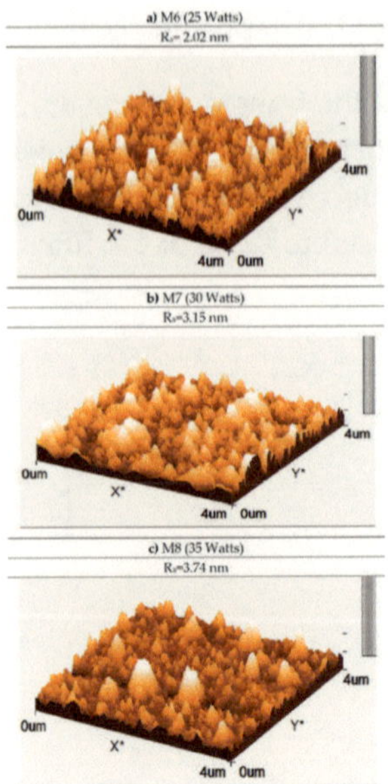

Figure 14. AFM three-dimensional images of μc-Si:H films as a function of the RF power: (a) M6 (25 W); R_a = 2.02 nm. (b) M7 (30 W); R_a = 3.15 nm. (c) M8 (35 W); R_a = 3.74 nm.

Figure 15. FE-SEM top view image of a μc-Si:H film.

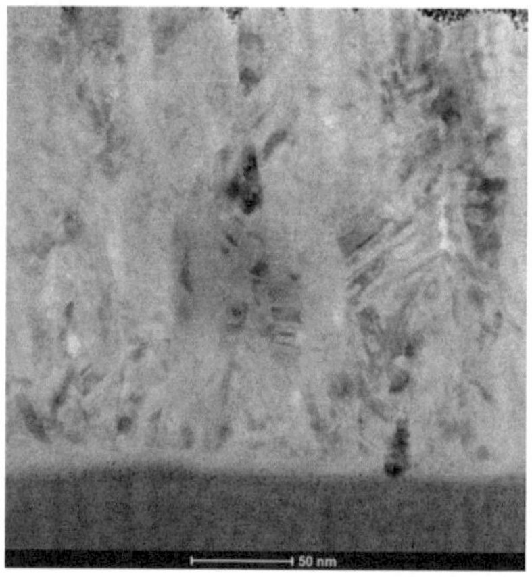

Figure 16. HRTEM transversal view of a μc-Si:H film.

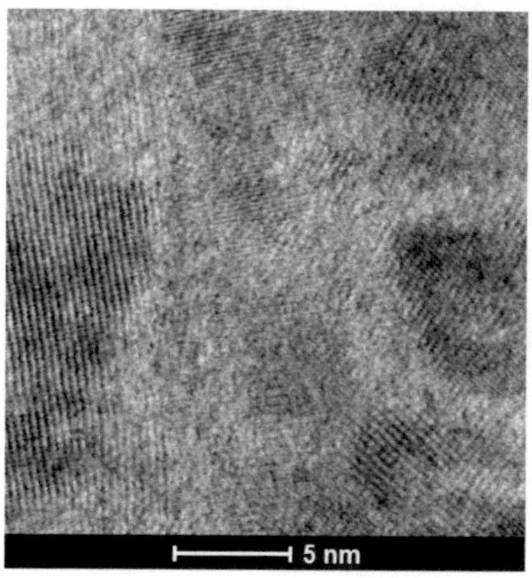

Figure 17. HRTEM transversal view of a μc-Si:H film.

5.2.3. Fourier transform infrared spectroscopy (FTIR) analysis

The IR absorbance spectra of the μc-Si:H thin films in the range 500–2200 cm^{-1} are shown in **Figure 18**. The total H_2 bonded content C_H of the films was calculated with the integration of the area under the Si—H mode (at 640 cm^{-1}) using Eq. (2), as in Section 4.2.3.

In **Figure 18**, the Si—H bonds in wagging and stretching modes are presented at 640 cm^{-1}, 700 cm^{-1}, 850 cm^{-1}, 920 cm^{-1}, and 2150 cm^{-1}, which are typical for μc-Si:H [29]. The band at 850–950 cm^{-1} observed in the samples corresponds to SiH_n bending modes; for this region low intensity is observed for the samples M7 and M8, which were deposited with larger RF power density (compared with the sample M6).

The band at 2000–2200 cm^{-1} is attributed to SiH, SiH_2, and SiH_3 stretching modes; the bands in this region are related to the presence of nanocrystals embedded in the a-Si:H matrix [41]. In **Table 6**, it is observed that the calculated H_2 content for the films of series #2 is in the range of 3–4.4%, which is low in comparison with similar work reported in literature [31]. The low H_2 content could be related to the fact that the excited Ar ion species that transfer their energy to the growing film and break the weak Si—Si bonds produced more ordered and relaxed network. In addition, large RF power density enhances the dissociation and ionization of precursor gases [35].

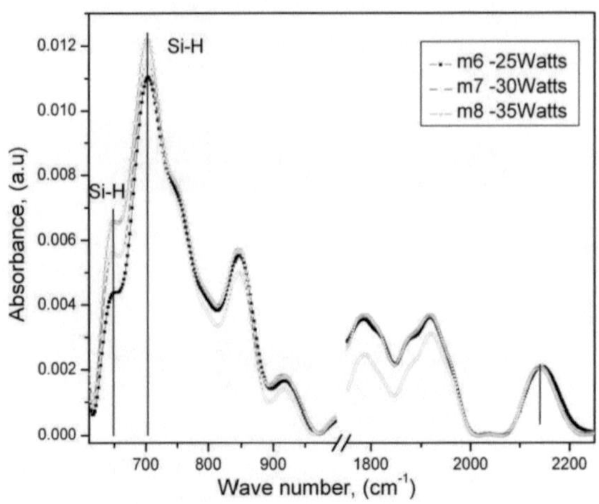

Figure 18. FTIR spectra for the μc-Si:H thin films.

Sample #	**R_a (nm)**	**C_H (%)**	**$X_{C\_(Raman)}$ (%)**
M6	2.02	2.9	67.9
M7	3.15	4.3	39.0
M8	3.74	4.4	55.8

Table 6. Surface roughness (R_a), H_2 content, and crystalline volume fraction for the deposited films.

http://dx.doi.org/10.5772/63522

5.2.4. Raman analysis

In **Figure 19** the Raman spectra of microcrystalline films with the presence of a peak related to a crystalline volume content are shown. The crystallinity in the samples M6–M8 is dominated by the peak at 520 cm^{-1} related to the transverse-optic (TO) mode, which is also observed for crystalline silicon [43].

In **Figure 19** an extra curve was introduced, deposited with the same deposition conditions of sample M6 but without Ar dilution, in order to observe the effect of Ar in the growth of μc-Si:H films. As is observed in **Figure 19**, that sample does not show the peak at 520 cm^{-1}, showing the importance of Ar in the dissociation of the SiH_4 molecules due to a larger ionization. In order to estimate the crystalline volume fraction ($X_{C(Raman)}$) of the films, a deconvolution in Raman peak intensities was performed using Eq. (3).

$$X_{C(Raman)} = \frac{I_c + I_m}{I_c + I_m + I_a} \tag{3}$$

where I_c is related to crystalline component at 520 cm^{-1}, I_m intermediate nanocrystalline component at 500–510 cm^{-1}, and I_a the amorphous phase at 480 cm^{-1} [43]. In **Table 6** the values of $X_{C\ (Raman)}$ obtained in the μc-Si:H films are shown.

As one can see in **Table 6**, there is no a lineal relation of X_C with respect to the increment of RF power density; the highest crystalline fraction in μc-Si:H films was found at RF power of 25 W. It is known that H_2 is present in Si—H bonds in a-Si:H deposited at low RF power and also that the H_2 content increases by increasing the RF power; in our films this behavior is presented. Therefore there is a compromise that suggests a suitable quantity of H_2 to grow μc-Si:H and a moderated RF power.

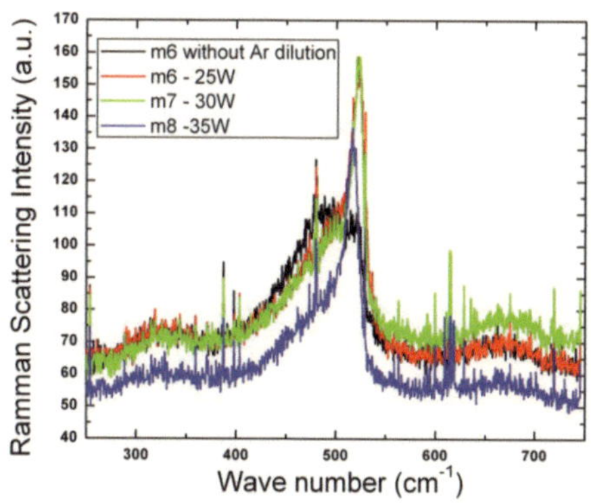

Figure 19. Raman spectra for the μc-Si:H thin films.

5.2.5. Optical properties

The optical band gap (E_{opt}) of the μc-Si:H films was calculated from transmittance measurements and the Tauc plot method. **Figure 20** shows the Tauc plots of the μc-Si:H thin films, where the RF power density was varied. For this set of samples, E_{opt} has no significant change. However the values of E_{opt} are in agreement with those reported in [33], where the μc-Si:H thin films have larger E_{opt} values than those of a-Si:H.

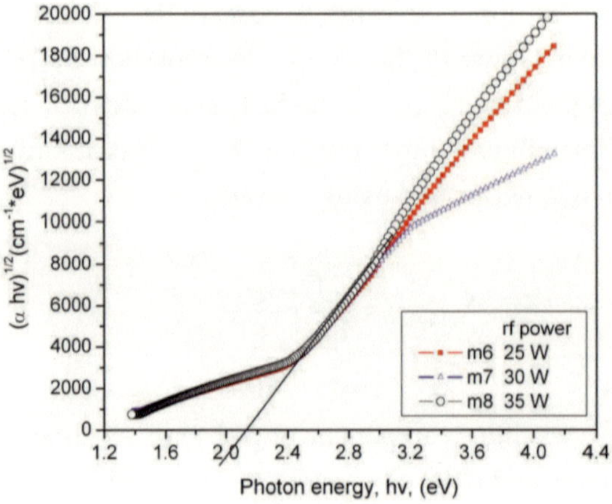

Figure 20. Optical band gap determination from the Tauc plots ($(\alpha h\nu)^{1/2}$ vs. photon energy) of μc-Si:H films as a function of the RF power.

5.2.6. Electrical properties

Temperature dependence of conductivity ($\sigma(T)$) measurements was performed in the μc-Si:H films; from that characterization the activation energy E_a was extracted, those values are shown in **Table 7**, and the E_a values are in the range of 0.15–0.27 eV, indicating doping, possibly related to oxygen contamination during deposition. Notice that the E_a values increase with an increment in the RF power.

As well the dark room temperature conductivity (σ_{dark}) and the photoconductivity (σ_{photo}) under AM 1.5 illumination conditions were measured, in the μc-Si:H films. The larger increment of σ from dark to AM 1.5 illumination in the μc-Si:H films is of about two orders of magnitude, measured in the film deposited with a RF power of 30 W (**Table 7**).

Sample #	Power (W)	V_d (A/s)	E_{opt} (eV)	σ_{dark} (Ω-cm)$^{-1}$	σ_{ph} (Ω-cm)$^{-1}$	E_a (eV)
M6	25	1.1	2.09	6.32E–7	1.21E–5	0.15
M7	30	0.9	2.1	1.59E–6	1.05E–4	0.15
M8	35	0.8	2.1	5.35E–6	7.98E–5	0.27

Table 7. Characteristics of μc-Si:H films.

Acknowledgements

We acknowledge to Fernando Mendoza, Diego García, and Daniel Phifer, from FEI, for their samples' preparation and HRTEM observation of the μc-Si:H films.

Author details

Mario Moreno[1*], Roberto Ambrosio[2], Arturo Torres[1], Alfonso Torres[1], Pedro Rosales[1], Adrián Itzmoyotl[1] and Miguel Domínguez[3]

*Address all correspondence to: mmoreno@inaoep.mx

1 National Institute of Astrophysics, Optics and Electronics, INAOE, Electronics Department, Puebla, Mexico

2 Electronics School, Meritorious Autonomous University of Puebla, BUAP, Puebla, Mexico

3 Research Center of Semiconductor Devices, Meritorious Autonomous University of Puebla, BUAP, Puebla, Mexico

References

[1] Roca i Cabarrocas P., Plasma enhanced chemical vapour deposition of amorphous polymorphous and microcrystalline silicon films, J. Non Cryst. Solids, 266–269, (2000) 31–37. doi: 10.1016/S0022-3093(99)00714-0.

[2] Bronner W., Kleider J.P., Bruggemann R., Mencaraglia D., Mehring M., Comparison of transport and defects properties in hydrogenated polymorphous and amorphous silicon, J. Non Cryst. Solids, 299–302, (2002) 551–555. doi: 10.1016/S0022-3093(01)01201-7.

[3] Roca i Cabarrocas P., Fontcuberta i Morral A., Lebib S., Poissant Y., Plasma production of nanocrystalline silicon thin films for large area electronic and devices, Pure Appl. Chemicals, 74 (3), (2002) 359–367. http://dx.doi.org/10.1351/pac200274030359.

[4] Brinza M., Adriaenssens G.J., Abramov A., Roca i Cabarrocas P., Influence of deposition parameters on hole mobility in polymorphous silicon, Thin Solid Films 515 (2007) 7504–7507. doi: 10.1016/j.tsf.2006.12.009

[5] Roca i Cabarrocas P., Fontcuberta i Morral A., Poissant Y., Growth and optoelectronic properties of polymorphous silicon thin films, Thin Solid Films, 403–404 (2002) 39–46. doi: 10.1016/S0040-6090(01)01656-X

[6] Fontcuberta i Morral A., Brenot R., Hamers E.A.G., Vanderhaghen R., Insitu investigation of polymorphous silicon deposition, J. Non Cryst. Solids 266–269 (2000) 48–53. doi: 10.1016/S0022-3093(99)00723-1

[7] Staebler D.L., Wronski C.R., Reversible conductivity changes in discharge-produced amorphous Si, Appl. Phys. Lett.31, 292–294 (1977). doi: 10.1063/1.89674

[8] Kim K.-H., Johnson E.V., Roca i Cabarrocas P., Irreversible light-induced degradation and stabilization of hydrogenated polymorphous silicon solar cells, Sol. Energy Mater. Sol. Cells, 105 (2012) 208–212. doi: 10.1016/j.solmat.2012.06.026.

[9] Cashmore J.S., Apolloni M., Braga A., Caglar O., Cervetto V., Fenner Y., Goldbach-Aschemann S., Goury C., Hötzel J.E., Iwahashi T., Kalas J., Kitamura M., Klindworth M., Kupich M., Leu G.-F., Lin J., Lindic M.-H., Losio P.A., Mates T., Matsunaga D., Mereu B., Nguyen X.-V., Psimoulis I., Ristau S., Roschek T., Salabas A., Salabas E.L., Sinicco I., Improved conversion efficiencies of thin-film silicon tandem (MICRO-MORPH™) photovoltaic modules, Sol. Energ. Mater. Sol. Cells, 144 (2016) 84–95. http://dx.doi.org/10.1016/j.solmat.2015.08.022.

[10] Van Sark W.G.J.H.M., Chapter 1 in thin films and nanostructures, vol. 30, In: Francombe M.H., ed., Advances in Plasma-Grown Hydrogenated Films, Academic Press, San Diego, 2002, 1–215.

[11] Bruno G., Capezzuto P., Madan A. eds., Plasma Deposition of Amorphous-based Materials. Academic Press Inc, San Diego, 1995.

[12] Cluster Tool for Prototyping [internet] 2016. Available from: http://www.mvsystemsinc.com/products/cluster-tool-for-prototyping/ [Accessed: 2016-05-01]

[13] Takatsuka H., Noda M., Yonekura Y., Takeuchi Y., Yamauchi Y., Development of high efficiency large-area silicon thin film modules using VHF-PECVD, Sol. Energy 77, 6, (2004), 951–960. doi: 10.1016/j.solener.2004.06.007

[14] Watanabe T., Tanaka M., Azuma K., Nakatani M., Sonobe T., Shimada T., Microwave-excited plasma CVD of a-Si:H films utilizing a hydrogen plasma stream or by direct excitation of silane, Jpn. J. Appl. Phys. 26, 8, (1987) 1215. http://dx.doi.org/10.1143/JJAP.26.1215.

[15] Hayama M., Kobayashi K., Kawamoto S., Miki H., Onishi Y., Characteristics of a-Si:H films deposited by electron cyclotron resonance plasma CVD, J. Non Cryst. Solids 97–98, (1987) 273–276. doi: 10.1016/0022-3093(87)90065-2

[16] McCormick C.S., Weber C.E., Abelson J.R., Davis G.A., Weiss R.E., Aebi V., Low temperature fabrication of amorphous silicon thin film transistors by dc reactive magnetron sputtering, J. Vac. Sci. Technol. A 15, (1997) 2770. http://dx.doi.org/10.1116/1.580821.

[17] Jadkar S.R., Sali J.V., Funde A.M., Ali Bakr N., Vidyasagar P.B., Hawaldar R.R., Amalnerkar D.P., Deposition of hydrogenated amorphous silicon (a-Si:H) films by hot-

wire chemical vapor deposition (HW-CVD) method: role of substrate temperature, Sol. Energ. Mater. Sol. Cells, 91, (2007) 714–720. doi: 10.1016/j.solmat.2006.12.009.

[18] Brinza M., Adriaenssens G. J., Electronic properties of hydrogenated amorphous silicon prepared in expanding thermal plasmas, J. Optoelectron. Adv. Mater. 7, (2005) 73–81.

[19] Roca i Cabarrocas P., Fontcuberta i Morral A., Lebib S., Poissant Y., Plasma production of nanocrystalline silicon particles and polymorphous silicon thin films for large-area electronic devices, Pure Appl. Chem., 74 (2002) 359–367. doi: 10.1.1.545.1422.

[20] Thomas Zimmermann, High-rate growth of hydrogenated amorphous and microcrystalline silicon for thin-film silicon solar cells using dynamic very high frequency plasma-enhanced chemical vapor deposition, Writings of the research center of Jülich Ltd., Vol. 183, 2013.

[21] Ambrosio R., Moreno M., Torres A., Carrillo A., Vivaldo I., Cosme I., Heredia A., Deposition and characterization of amorphous silicon with embedded nanocrystals and microcrystalline silicon for thin film solar cells, J. Alloy Compd., 643, 1, (2015) S27–S32. doi: 10.1016/j.jallcom.2014.11.105.

[22] Moreno M., Torres A., Ambrosio R., Zuñiga C., Torres-Rios A., Monfil K., Rosales P., Itzmoyotl A., Study of the effect of the deposition parameters on the structural, electric and optical characteristics of polymorphous silicon films prepared by low frequency PECVD, Mater. Sci. Eng. B: Adv. 176 (2011) 1373–1377. doi: 10.1016/j.mseb.2011.01.022.

[23] Huang T.-C., Huang J.-L., ChengTim K.-T., Robust circuit design for flexible electronics, IEEE Des. Test Comput., 28, (2011) 8–15. doi: 10.1109/MDT.2011.74.

[24] Stutzmann M., Defect density in amorphous silicon, Philos. Mag. B: Phys. Condens. Matter 60, (1989) 531–546. doi: 10.1080/13642818908205926.

[25] Springer Handbook of Electronic and Photonic Materials, In: Kasap S., Capper P. eds., Springer Science+Business Media, Inc., New York, 2006.

[26] Deng X., Schiff E.A., Amorphous silicon based solar cells, In: Luque A., Hegedus S., eds., Handbook of Photovoltaic Science and Engineering, John Wiley & Sons, Chichester, 2003, 505–565.

[27] Matsumura H., Nakagome Y., Furukawa S., Conductivity and p–n type control of fluorinated amorphous silicon (a-Si:F) without incorporating hydrogen, J. Appl. Phys., 52, (1981), 291. http://dx.doi.org/10.1063/1.328491.

[28] Singh J., Shimakawa K., eds., Advances in Amorphous Semiconductors, Taylor & Francis, New York, 2003.

[29] Kroll U., Meier J., Shah A., Hydrogen in amorphous and microcrystalline silicon films prepared by hydrogen dilution, J. Appl. Phys. 80 (1996), 4971–4975. doi: 10.1063/1.363541. http://dx.doi.org/10.1063/1.363541.

[30] Lucovsky G., Vibrational spectroscopy of hydrogenated amorphous silicon alloys, Sol. Cells 2 (1980) 431.

[31] Lebib S., Roca iCabarrocas P., Structure and hydrogen bonding in plasma deposited polymorphous silicon thin films, Eur. Phys. J. Appl. Phys. 26 (2004) 17–27. http://dx.doi.org/10.1051/epjap:2004018.

[32] Guo L., Ding J., Yang J., Cheng G., Ling Z., Yuan N., Effects of high hydrogen dilution ratio on optical properties of hydrogenated nanocrystalline silicon thin films, Appl. Surf. Sci. 257 (2011) 9840–9845, doi: 10.1016/j.apsusc.2011.06.038

[33] Li W., Xia D., Wang H., Zhao X., Hydrogenated nanocrystalline silicon thinfilm prepared by RF-PECVD at high pressure. J. Non Cryst. Solids 356 (2010) 2552–2556. doi: 10.1016/j.jnoncrysol.2010.07.064

[34] Roca iCabarrocas P., Kim K.H., Cariou R., Labrune M., Johnson E.V., Moreno M., Torres Rios A., Abolmasov S., Kasoui S., Low temperature plasma synthesis of nanocrystals and their application to the growth of crystalline silicon and germanium thin films, Mater. Res. Soc. Symp. Proc. 1426 (2012). doi: 10.1557/opl.2012.1094

[35] Aguas H., Roca iCabarrocas P., Lebib S., Silva V., Fortunato E., Martins R., Polymorphous silicon deposited in large area reactor at 13 and 27 MHz, Thin Solid Films. 427 (2003) 6–10. doi: 10.1016/S0040-6090(02)01172-0.

[36] Amor S.B., Atyaoui M., Bousbih R., Ouertani R., Dimassi W., Ezzaouia H., Effect of substrate temperature on microstructure and optical properties of hydrogenated nanocrystalline Si thin films grown by plasma enhanced chemical vapor deposition, Sol. Energ. 103 (2014) 12–18. http://dx.doi.org/10.1016/j.solener.2014.06.024.

[37] Badran R.I., Al-Hazmi F.S., Al-Heniti S., Al-Ghamdi A.A., Li J., Xiong S., A study of optical properties of hydrogenated microcrystalline silicon films prepared by plasma enhanced chemical vapor deposition technique at different conditions of excited power and pressure, Vacuum 83 (2009), 1023–1030. doi: 10.1016/j.vacuum.2009.01.009.

[38] Shah A., Vallat-Sauvain E., Torres P., Meier J., Kroll U., Hof C., Droz C., Goerlitzer M., Wyrsch N., Vanecek M., Intrinsic microcrystalline silicon (mc-Si:H) deposited by VHF-GD (very high frequency-glow discharge): a new material for photovoltaics and optoelectronics, Mater. Sci. Eng. B 69–70 (2000) 219–226. doi: 10.1016/S0921-5107(99)00299-8

[39] Moreno M., Boubekri R., Roca P., Cabarrocas I., Study of the effect of different fraction of large grains μc-Si:H:F films on the infrared absorption on thin film solar cells, Sol. Energ. Mater. Sol. Cells 100 (2012) 16. doi: 10.1016/j.solmat.2011.05.030.

[40] Matsui T., Kondo M., Matsuda A., Origin of the improved performance of high-deposition-rate microcrystalline silicon solar cells by high-pressure glow discharge, Jpn. J. Appl. Phys. 42 (2003), L901–L903. http://dx.doi.org/10.1143/JJAP.42.L901.

[41] Tang Z., Wang W., Zhou B., Wang D., Peng S., He D., The influence of $H_2/(H_2 + Ar)$ ratio on microstructure and optoelectronic properties of microcrystalline silicon films

deposited by plasma-enhanced CVD, Appl. Surf. Sci. 255 (2009), 8867–8873. doi: 10.1016/j.apsusc.2009.06.074

[42] Roca iCabarrocas P., Chaabane N., Kharchenko A.V., Tchakarov S., Polymorphous silicon thin films produced in dusty plasma: applications to solar cells, Plasma Phys. Control. Fusion 46 (2004) B235–B243. doi: 10.1088/0741-3335/46/12B/020.

[43] Kaneko T., Onisawa K., Wakagi M., Kita Y., Minemura T., Crystalline fraction of microcrystalline silicon films prepared by plasma enhanced chemical vapor deposition using pulsed silane flow, Jpn. J. Appl. Phys. 32 (11A), (1993), 4907–4911. doi: http://dx.doi.org/10.1143/JJAP.32.4907.